THE 100

Greatest Combat Pistols

Hands-On Tests and Evaluations of Handguns from Around the World

Timothy J. Mullin

Paladin Press
Boulder, Colorado

3/9/95

To John Rayfield

with best wishes

of the

author

To my wife, Eleanor, for understanding that Sundays are for testing weapons or going to gun shows, and to my daughter, Catherine, who is a blessing.

The 100 Greatest Combat Pistols:
 Hands-On Tests and Evaluations of Handguns from Around the World
by Timothy J. Mullin

Copyright © 1994 by Timothy J. Mullin

ISBN 0-87364-781-5
Printed in the United States of America

Published by Paladin Press, a division of
Paladin Enterprises, Inc., P.O. Box 1307,
Boulder, Colorado 80306, USA.
(303) 443-7250

Direct inquiries and/or orders to the above address.

Table of Contents

Foreword

TJ. Mullin and I have been friends and shooting companions for many years. A frequent pastime we have indulged in over that period is speculation about what handgun we would have chosen at various points throughout history for various purposes. As both of us had received training as infantry officers, our discussions frequently revolved around the merits of various automatics and revolvers as armament for the infantryman. Staff officers and many high-ranking line officers rarely appreciate the handgun as a combat weapon at all, but grunts—including those with one or two bars on their shoulders and occasionally those with oak leaves—realize that the pistol remains the foot soldier's last resort. It must be remembered, too, that rotor or fast-mover pilots, armored crewman, artillerymen, and even those senior or staff officers who denigrate the handgun may, depending upon circumstances, end up as foot soldiers, and, for all but the infantryman, their pistols may be their primary weapons.

As our discussions about the merits of various contemporary weapons progressed, we also began to argue the merits of the weapons available to our fathers, grandfathers, and great grandfathers. T.J. just decided to take our shooting discussions to the next logical step. He developed a test program based on the necessity for a battle

pistol to work reliably and to grant the ability to successfully stop an enemy fighting man quickly at battle ranges. Even the "fog of war" was simulated to some extent with the cinema range tests carried out on each of the pistols to determine their handling characteristics under stress.

A secondary purpose of the tests was to provide the reader a little more information that might prove highly useful should he or she some day find himself/herself in danger with only a "Ramadan night special" available at the local bazaar. Various instances have shown me that wide familiarity with a variety of weapons may prove invaluable, even life saving. More than once, I've found myself on assignment in a foreign country where my choice of weapons was determined either by what my hosts had provided for me or by what I could rapidly scrounge on my own. As a result, I once found myself guarding a very high-threat target—what a colleague of mine calls a "threat level ten"—one who was being actively stalked by an honest-to-Allah hit team, and my armament was a Savage .32 auto—one I should note I was glad to have, as it was a giant step up from the Spanish Special Forces knife I was backing it up with. Another time, I found myself in the midst of an insurgency armed with a Webley .380 Mk IV revolver and eight rounds of ammunition. A friend of mine, a former MI6 agent, can go me one better. At one point during a rebellion in Africa, he was armed with a Montenegrin Gasser revolver and a handful of old black powder cartridges.

I guess the point I'm trying to make is that the best combat handgun is the one you have in your hand when someone is shooting at you. But if it happens to be one you've just taken off a dead enemy (or a friend for that matter), it's useful only if you have some idea how it works. I have given foreign weapons familiarization courses to members of some government agencies specifically to prepare them for this contingency. This book is another step toward such preparation, encompassing as it does many of the handguns most likely to be available during times of strife. Older war weapons have a way of turning up again in times of conflict; take a look at the arms in the hands of some of those fighting in Bosnia. I would be astonished if old Balkan favorites such as the Rast & Gasser or Roth-Steyr aren't still accounting for those who either want to carry out "ethnic cleansing" or those who choose not to be "cleansed."

T.J. Mullin has put in many hundreds of hours on this work, and I'm pleased to have taken part in most of the tests. When we set out to assemble the weapons necessary for this book, we felt that certain ones would prove quite difficult while others would prove relatively easy. Seven years down the road, I'm amazed at some of the guns that have been tested for this work. Friends who collect esoteric military weapons have been con-

tacted, favors from those in intelligence agencies, special operations armories, and police ballistic labs have been called in, and many dollars have been spent at gun shows to purchase guns necessary for the tests. Numerous preconceptions have been destroyed. Guns that have always looked ungainly in photographs have proven decades ahead of their time and worthy of consideration, even today, as combat handguns. Other handguns, long considered classic, have proven to be very poor for real combat use.

Whether the guns tested were borrowed or purchased, though, it is highly likely that this book could only have been written in the United States, where the Second Amendment is an actuality, where one doesn't have to be a secret policeman or a museum curator to have access to the handguns that have shaped the fabric of our century.

Take my word; you're in for an enjoyable and informative reading experience. This is one gun book that you'll read, then probably reread individual evaluations again and again, especially as you read tales or view films of far-off places and far-off wars. If you're like me, you'll want to look up T.J. Mullin's evaluations and see if that Webley-Fosberry that killed Sam Spade's partner really was an "automatic revolver," or if the Walther PPK really merited the affection James Bond held it in.

Leroy Thompson
St. Louis, MO
1994

Why There Is a Need for This Book

There are many books on the subject of handguns, so why do another one? Most of the previously published works are reference or picture books showing how to detail-strip weapons. This book is different. It is a hands-on, practical comparison of more than 100 handguns from around the world that have been used for the past 100 years.

In order to see how each weapon performed on the target range and under simulated combat conditions, I shot *every* weapon in this book. Based on these results, readers will be able to judge how well individual handguns function and how they actually shoot and respond in combat situations. It is ideal for those who are now or may be stationed in remote environs and have only a limited selection of weapons available. After reading this book, the intelligence agent stationed overseas, the missionary, the professional bodyguard, the soldier, or the traveling citizen/ businessman will understand the relative merits of various weapons that he might happen across and the proper ammo to use under less-than-ideal conditions. Even though his choice of weapons may be limited, this individual will know his options and be able to select the best handgun for his particular purpose from what is available. And he will know how it can be made workable.

But *The 100 Greatest Combat Pistols* is not just for expatriates or people who crisscross the globe. It is also an invaluable resource for weapons aficionados and collectors who might not be able to get their hands on some of these rare, exotic, or historic handguns—some of which cannot be found outside museums or special operations arsenals.

A WORD ABOUT THE WEAPONS TESTED

Each of the weapons tested in this report was a common example of the weapons used by the armies in its day. No special specimens were selected. Some of the handguns were in good shape; some were in excellent shape. All of them were good, honest-firing models, not modified in any way to improve their accuracy or speed of handling. Many times, I would have preferred to modify them in some fashion, at the very minimum to paint the sights. However, to keep the tests consistent, it was important that they be unmodified examples of generally available military weapons.

TESTING METHODS

For the comparisons to have any validity, I had to test each weapon and evaluate it based on its accuracy, reliability, and availability.

Each weapon was given a twofold test. First, using the two-hand Weaver stance position, I fired each weapon at 50 feet off-hand on a bull's-eye range. Second, I tested them on a simulated combat range—in this case, a cinema range. The cinema method involves projecting specific scenarios on a screen and having the shooter respond. The films that I used were police simulations involving innocent bystanders and moving targets, all requiring a different level of lethal force.

It is very important to use both target and simulated-combat testing methods because a weapon that is fine on the target range may be terrible on the cinema range—or in a real-life scenario—and vice versa.

It is also important to test the weapons against a known standard. For a reference medium, I used the Model 19 four-inch barrel Smith & Wesson .357 Magnum loaded with 148-grain wadcutters. This is an easily obtainable, standardized weapon with target sights. It is similar to what is commonly available throughout the Western world, and it is familiar to most people who read this book. Thus you, the reader, can evaluate the performance of each weapon even though you may never have fired the weapon being tested.

AMMUNITION

Another factor that greatly influences firearms tests is the ammunition used. Ammunition for these tests was the same type within a given caliber. All full-metal-jacketed ammunition of commercial manufacture was used for autoloaders. No hand loads were used except in obsolete and rare calibers that were unavailable commercially. I used Fiocchi, Winchester Western, and Remington Arms ammunition. All were full-metal-jacketed loads or otherwise military-standard loading except as noted. I used no special ammunition that would aid accuracy or reliability because I wanted to produce the same results the soldier would encounter in the field.

CONSIDERATIONS FOR SELECTING MILITARY WEAPONS

To properly evaluate military weapons, it is important to understand some of their historical aspects and uses. In the early days, only officers and noncommissioned officers used handguns. Their job was to direct troops, using their handguns only in emergencies.

The role of officers—and the weapons available to them—has changed greatly over the years, so that now officers at the company level are typically armed with shoulder weapons as well. For instance, during World War II, a captain in the military forces probably carried only a handgun. Nowadays, a captain in the U.S. Army carries an M16 rifle in addition to a handgun. He is also laden with magazine pouches, grenades, claymores, and other equipment. The ability to carry a handgun has been severely limited by the adoption of rifles. Many of the weapons we are going to test in this book were designed to be carried on the hip, and in today's army it may be difficult for infantrymen to carry such weapons.

Of course, there are always artillerymen, pilots, and transportation corps types who are not on the battle line but who need to be armed and prepared to defend themselves. They carry—and will continue to carry—handguns because rifles are unwieldy or inconvenient for them. We may see a trend in the future for militaries to adopt one weapon for infantry people and another for these second-level troops, as was done by U.S. forces in World War II with the .30 carbine.

It may well be that the weapon carried by the infantry is less efficient—but smaller and easier to carry—than that carried by other personnel. For example, when I was an infantry officer, I carried a Model 60 .38 Special Smith & Wesson stainless steel revolver in the upper left-hand breast pocket of my fatigues. It was out of the way of my magazine pouches and rucksack straps and protected from mud and crud, and it allowed ready access to my handgun. I did not kid myself into thinking that a Model 60 2-inch was a serious fighting handgun, but it was a great aid to an M14 or M16 that could get misplaced or become inoperable. On the other hand, if I were armed with only a handgun—as might be the case in a back-of-the-lines area or as an advisor (not leading troops into combat like many of our "advisors" in the Vietnamese army)—then I would choose a serious fighting handgun, such as a Glock 17 or SIG P220 in .45 ACP.

How to carry your weapon also becomes critical in a combat situation. Typically, infantry officers do not carry their handguns in holsters on their right hips anymore. They carry them in shoulder holsters, fatigue jacket pockets, or upper breast shirt pockets. Method of carry is critical when considering what is an appropriate military weapon.

The Hague Conference of 1907

Another factor in the choice of military weapons is the Hague Peace Conference of 1907, where the use of expanding ammunition in military operations was banned, thus limiting the power available in a given handgun. In

police circles, for instance, no one would dream of using a 9x19mm pistol with full-metal-jacketed loads: they are considered seriously deficient in stopping ability. Military forces, on the other hand, are compelled to use ammunition that complies with that convention; so when we evaluate military weapons, we must take this into consideration.

Ballistic Vests and Helmets

A more recent consideration in the selection of military handguns is the introduction of ballistic vests and helmets that can stop rifle fire at 10 meters and pistol ammunition at point-blank range. Military loads that previously were perfectly adequate no longer are. To date, this has not become a major problem for

Front view of 1990s era U.S. soldier with ammunition pouches, body armor, ballistic helmet, and goggles, plus weapon, which will stop bullets. You need a lot of penetration now, unlike earlier times.

Rear view of current U.S. soldier. Note the ballistic vest and helmet and the ammunition pouches.

militaries because the major actions have been basically low-technology Third World affairs, or they have involved situations where technologically superior forces are faced with climatic conditions that render ballistic vests difficult, but not impossible, to use at all times. For example, in Vietnam, the high humidity made the vests extremely uncomfortable, and comfort is not the only limiting consideration for Kevlar vests in the tropics: when wet, Kevlar loses a significant portion of its ballistic resistance, which it regains upon drying.

However, as ballistic vest technology improves, making vests lighter and cooler to wear and more impervious to environmental conditions, militaries may be forced to change completely the handguns and loads used. We are on the cutting edge of this area. One solution may be the general adoption or loads such as the French THV or Arcane, hypervelocity loads of solid material that comply with the convention yet are still capable of being fired in a handgun and possibly penetrating ballistic vests and helmets. Consequently, previously ineffective weapons, such as .32 ACP autoloaders, may turn out to be quite effective when loaded properly. Though it is difficult to imagine this happening, the results of my testing of the French THV rounds on ballistic material were very impressive.

Ease of Manufacture

In today's world, it is critical that military handguns be manufactured easily and cheaply. They should require minimal machine time and strategic resources. The availability of inexpensive, easily produced weapons has no direct bearing upon their performance in the military theater, but it is certainly is a concern for military procurement specialists.

Safety

As anyone who is familiar with police operations knows, you find that many police officers "trained" in the use of handguns are still dangerous. In the military, with inadequately trained troops, this is even more the case. Many people in the military have only a nodding acquaintance with firearms before they get there, if at all, and no interest whatsoever in handguns. They get very little in the way of training with handguns. For instance, during the World War II, it was not uncommon for German soldiers armed with handguns to be given 25 rounds, which were supposed to last them throughout the entire war. So it is in most situations of that type. Thus, it is critically important for military planners to pick a weapon that is safe to handle. Many times they trade efficiency for safety.

In the past, the safety problem was circumvented by continuing to use weapons of questionable safety and then selecting a carry system that restricted access to the weapon. Rather than training their forces to carry the .45 Government Model cocked and locked, which is the way it was designed to be used, militaries ordered soldiers to carry it with the chamber empty, hammer down—clearly restricting the soldiers' abilities to respond quickly. It is interesting to note that U.S. military manuals prior to and during World War I dictated that the .45 Government Model be carried cocked and locked with the side safety on. It was only later that the prescribed method was the hammer-down, chamber-empty method. Even U.S. military police units are told to carry their weapons in that fashion: hammer down, chamber empty. To anyone familiar with handguns, it seems foolish, but military forces typically are not overly familiar with handguns.

Acknowledgments

I truly believe that this book represents new ground in the gun field. Many handgun books have preceded it, but most of them have been either dry reference books, such as *Small Arms of the World*, or picture books. *The 100 Greatest Combat Handguns* is the only book to provide you with the results of actual tests of the handguns being used as they were intended: in combat situations. In hands-on tests, I compare each handgun with something you will be able to relate to and provide an informed evaluation of each.

I do not think that this book could have been completed anywhere else other than the United States, at least not by a private person like me who is not a curator of a national museum. In a few other countries—Switzerland, for example—various handguns are available to the general populace, but they do not have the variety of weapons that we do here. Thus, my first thanks must go to our Founding Fathers, who were farsighted enough to include in the Constitution a prohibition against government infringement on our right to own weapons. Next, I give thanks to all those brave men and women who have fought so hard since 1789 to preserve those liberties.

On a more personal note, I wish to thank my friend Leroy Thompson, who took photos, assembled weapons, and discussed this topic in general with me. For a long time, Leroy and I were the only ones who thought this project was useful or interesting.

I would also like to thank Tom Knox for helping to assemble some of the tested weapons; Joe Davis for allowing me to shoot his British collection; Richard Hoffman, who loaned me his French and German ordnance revolvers; Shawn McCarver for his assistance; Ed Seyffert for his aid; and Dave Noll, who rendered yeoman service in locating some of the oddities needed to complete this book.

The Need for a Military Handgun

When firearms were first developed, the difference between handguns and shoulder weapons was quite narrow, if it existed at all. A firearm was held with one hand, and the other held the burning fuse.

About as soon as guns were developed, Europeans quickly fielded a matchlock hand cannon that rested on a pivot mount on the front of the saddle. The "stock" was a rod that rested against a ring on the cavalryman's chest, and the slow-match was held in the right hand. The horse just took care of itself. Since the muzzle of the gun was behind the horse's ears, it must have been fun. That's probably why they developed smaller one-hand weapons. The term pistol is derived from the Italian word for horse soldier.

The revolver popularized by Samuel Colt in the 1830s and distributed widely starting in the 1840s changed the relationship between the handgun and the shoulder weapon. Whereas previously the weapons were of equal firepower, the revolver allowed the cavalryman to fire a number of times prior to reloading. Many Civil War-era cavalrymen carried as many as six revolvers on their horses, firing each in turn until it was empty. This permitted the cavalryman to fire more than 30 times quickly, a tremendous increase in firepower over the single-shot horse pistol of a few years earlier. These weapons were so efficient that they displaced the sword as a practical weapon.

Infantry troops also used the handgun with great relish during this period. In European armies, private ownership of firearms was not common among the enlisted men, but officers typically purchased their own weapons and still commonly used swords. In the United States during this period, the fighting handgun was appreciated by the men of the day, both civilian and military. In the military, the cavalry officer of the American West developed great skill with the heavy handgun, including learning to use it at great range. More than a few officers learned that the 7 1/2-inch .45 Colt revolver could be used to hold hostile Indians well beyond range of their bows or smoothbore muzzle loaders.

During World War I, trench warfare quickly pointed out the need for a handgun capable of stopping an enemy at close range in a tight spot. Although European armies reserved handguns for officers and senior noncommissioned officers, the conditions at the front soon resulted in an urgent demand by all ranks for handguns. In the French army, demand was so high that procurement officers resorted to purchasing hundreds of thousands of .32 ACP self-loader pistols. Hardly the answer to the trench warriors' prayer for an effective fighting handgun, but it was better than a club or bayonet.

When World War I ended, the various armies of the world found themselves with many thousands of redundant handguns. Obviously, a need to rationalize the system existed, but militaries had no money to spend on such a low-priority task. I suspect that many high-grade officers found it quite alarming that individual soldiers had such a small, easily concealed weapon available, preferring instead that they alone be armed in such a manner. Other than in times of war, military organizations are elitist by definition, with people getting their authority and position by fiat rather than by ability or expertise.

World War II found the military forces with few handguns available to meet the tremendous demand. Further, the new equipment that the men had to carry limited their ability to carry an additional handgun. But use of specialized troops created a new demand, as did the surprise attacks by partisan guerrilla forces or elite military units that operated far behind the front. In the German army, for instance, individual soldiers were issued a pistol and 25 rounds of ammunition as some means of protection since they frequently lived in extremely hostile environments in occupied countries.

Although fighting troops did carry handguns, the rise of the submachine gun and M-1 carbine did change the role of handguns from that in World War I.

When the atomic bomb blasts at Hiroshima and Nagasaki brought a sudden end to the war, many military planners thought the fighting infantry had been replaced by the missile and bomb. By the mid-1950s, for instance, only one handgun per battalion was authorized in the British army. This theory proved shortsighted with the arrival of the various wars of national liberation and guerrilla campaigns that began during the late 1940s and continue through the present.

Individual soldiers are always making heavy demands on whatever handguns are available. Soldiers from various armies buy them and smuggle them to the front, "liberate" them from the enemy, and steal them from their own supply sources. Equally vigorous are the attempts by senior military planners to take them away from individual soldiers by telling everyone who will listen (and only inexperienced people will listen) that pistols are heavy and dangerous and not fit for soldiers.

However, anyone who has ever been in a foxhole at night with the enemy at his front or lived among hostile populations soon learns to cherish any handgun available. During the Vietnam War, all sorts of handguns, many of dubious quality, were seen in the hands of U.S. troops who, according to their Table of Organization and Equipment (TO&E), had no need for a handgun. Shake a GI upside down and frequently out falls a spare autoloader or revolver, something to give him a little edge in an emergency.

The modern military man may very well have a selective-fire assault rifle, claymore mines, grenades, and helicopter gunships, support aircraft, and artillery on call, but as anyone who has ever worn web gear knows, in the end, only *you* can prevent your being harmed. Only *your* skill with *your* weapons will suffice. And if you become separated from your rifle or cannot carry it because of your duties, or if your rifle malfunctions, then you have only your handgun to get you through. No number of planners or logistical arguments can alter that fact. Thus, as long as men meet in combat, the need for a military handgun will exist.

Mannlicher
M1900/M1901

7.63x27.5mm
(7.62, 7.65 Mannlicher)

The Mannlicher M1901 was one of the first auto-loading pistols found in modern armies. It reflects a period when pistol designers were unclear about how a pistol should be loaded and what caliber worked best.

The pistol is well made out of good material, and its grip allows the knuckles to line up rather than be staggered across the grip—the sign of a good grip in my estimation. Although the barleycorn front sight made it difficult to shoot easily on the formal range, it did shoot accurately, achieving a 1 15/16-inch group. The front sight is difficult for formal target work, but is very fast on the darkened cinema range. The rear sight is lower and has a very narrow notch that makes it difficult to pick up rapidly. The fact that the safety and hammer are both light-colored and that the rear sight is dark makes for a cluttered appearance and requires target-acquisition time. The safety is very slow to disengage and requires you to break your grip or carry it hammer down. If you do that, you need to break your grip to cock it. A particularly bad feature is that the recoil causes the safety to flip on, preventing you from firing a second shot. The hammer would drop, but the gun would not fire because the safety was blocking the hammer.

The cartridge is also obviously very light in stopping power: always a flaw in a military handgun. Because of the stripper-clip reloading feature, the weapon must be shot out before it can be reloaded. The M-1 Garand shares this same problem; a lot of people don't mind, but I have always shunned the M-1 rifle for that very reason. On a pistol that has a lot less stopping power and is likely to be used at closer range than a .30-06 rifle, the feature is even worse. When this pistol was designed, not all designers had concluded that self-loading pistols should have detachable magazines; the Roth-

SPECIFICATIONS

Name: Mannlicher M1901

Caliber: 7.63x27.5mm

Weight: 1.8 lbs.

Length: 8 3/4 in.

Feed: In-grip fixed magazine

Operation: Blowback with decelerating devices

Sights: Front blade; rear notch

Muzzle velocity: NA

Manufacturer: Steyr

Status: Obsolete

The author testing the Mannlicher M1901 handgun.

Right side of Mannlicher M1901.

Photo from rear of grip on Mannlicher M1901. The grip feels good in the hand.

The fine grip of the M1901 is illustrated here. Note how the shooter's knuckles line up.

The safety system on the Mannlicher M1901 is difficult to engage rapidly.

The safety is applied on this M1901.

Steyr 07 and the Mauser M96 suffer a similar disadvantage. Still, both the Roth-Steyr and the Mauser were faster on the cinema range than the Mannlicher M1901 and are preferable to it.

Most of the weapons of this type that you encounter are ex-Argentine army weapons. The army adopted the weapons to arm its cavalry units. This seems like a poor choice for a variety of reasons: reloading must have been difficult off the back of a stallion, the caliber was so light that you could not be assured of dropping your adversary, and you always had to worry about whether the safety would flip on by mistake. Remember these shortcomings if some better weapon is available.

Even though in a hammer-down position, the applied safety prevents the hammer from resting on the firing pin.

Even in the hammer-down position, the safety applied prevents the hammer from resting on the firing pin.

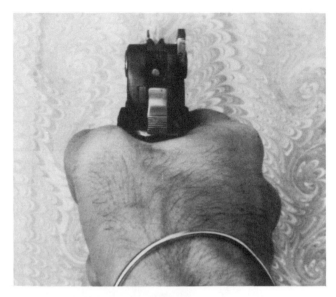

View of applied safety and sights.

Rast & Gasser
M1898 Revolver

8x27mmR
(8mm Gasser)

This revolver was the last Austrian revolver manufactured to arm the empire of Franz Joseph. No doubt these weapons are still found in drawers and under beds throughout Central Europe and the Balkans. Certainly, given the low pressure of the loads and the rugged construction of the weapons, it is unlikely they would ever wear out.

Recoil in this weapon was quite low, evidence of an underpowered load. It must be recalled, of course, that this was the period when rifles were being reduced in bore diameter without losing effectiveness, and it would take World War I to make it clear that a similar situation did not exist with handguns. The gun I tested was in excellent condition, with all small parts numbered to the weapon, revealing great care in its construction. The trigger was smooth faced and of a size almost identical to the "combat trigger" found on the modern Smith & Wesson

SPECIFICATIONS

Name: Rast & Gasser M1898

Caliber: 8x27mmR

Weight: NA

Length: NA

Feed: Revolver

Operation: Double action (DA)

Sights: NA

Muzzle velocity: NA

Manufacturer: Rast & Gasser

Status: Obsolete

revolvers. Although the reloading was quite slow, the system should be compared to that found on the contemporary Colt Single-Action Army. The grips were quite straight, and although this did interfere with instinctive pointing ability, they were shaped so as to allow the finger to rest on the trigger at the correct point for good double-action work. The firing pin was not affixed to the hammer, as is common even today, but instead was integral with the frame, reminding me of the current Colt Python. I cannot imagine why they went to this expense, but it was a nice touch. Cylinder capacity was eight rounds, and the power level of the revolver cartridge was similar to that found in the autoloading Roth-Steyr 1907 pistol. No doubt the troops of the period thought the loading system of the Roth-Steyr 07 was much superior to that found in the Rast & Gasser revolver, even though today we regret the absence of a detachable magazine.

The front sight on this revolver was very small and hard to see on both the formal range and the cinema range. The rear sight was also hard to use because of its size and width. One very nice feature on this weapon was that the side plate swings out for easy cleaning. If you have ever dropped your revolver in a mud hole or in seawater, you'd like this feature. Revolver designers today (don't laugh—look at the new Manurain!) would do well to copy this feature. In addition to simplifying cleaning, it also makes replacing parts easier. My only fear would be since it is so easy to open, the troops would be opening them all the time and playing with the parts. I think I could live with that prospect to get the ease of cleaning and replacing of parts that such a swinging side plate allows.

All in all, this is an interesting design, but it is limited by the caliber and slow reloading. I know if I were an Austrian officer, I would have quickly dumped my Rast & Gasser for a Roth-Steyr 07 as soon as I could get one. However, if you compare this with other rifle-bored ordnance revolvers of the pre-1914 era, the Rast & Gasser comes out ahead in a couple of areas and is no worse than others in the remaining ones.

The author field-shooting the Rast & Gasser M1898 8mm revolver on a cold day (right).

Right side of M1898 Rast & Gasser (lower right).

The smooth combat trigger on M1898 Rast & Gasser would be acceptable today (below).

Rast & Gasser M1898 Revolver

Left side of M1898 Rast & Gasser.

As is common with many European ordnance revolvers, the M1898 opens up without tools to allow the soldier to clean the weapon more easily. This is a very desirable feature on a combat revolver.

The 50-foot test target shot with the M1898 Rast & Gasser 8mm.

Roth-Steyr M1907
8x18.5mm
(8mm Roth-Steyr)

Invented by Czech arms designer Karel Krnka, the "Roth"-Steyr was the first officially adopted autoloading handgun of Austria-Hungary.

The first thing that strikes you about this pistol is how beautifully made and finished it is. My specimen was made in 1910 at the Steyr factory in Austria and was no doubt issued to a Hungarian or Austrian cavalry unit. It was much better finished or made than anything you find coming out of factories today, except a SIG P210-1. Disassembly shows many complicated machine cuts that could be duplicated today only at great expense. A lot of work went into the making of this pistol, and a lot of thought went into its design.

Firing on the formal range was complicated by the trigger system that made shooting more difficult than with a standard trigger system. If one used the obsolete off-hand single-hand firing platform, this difficulty would be even greater; no

SPECIFICATIONS

Name: Roth-Steyr M1907

Caliber: 8mm

Weight: NA

Length: 9.1 in.

Feed: 10-rd., in-line, nonremovable mag.

Operation: Recoil

Sights: Blade with notch

Muzzle velocity: 1,045 fps

Manufacturer: Steyr

Status: Obsolete

doubt its trigger system, which is such an asset for real-world use, is what caused target shooters to shun the weapon.

The 8mm Steyr ammunition is not to be confused with the .32 ACP cartridge, even though the rounds are similar in size. The pistol uses a locked breach, and the loading is more akin to the .30 Luger than .32 ACP. The bullet is faster than .32 ACP, going about 1,100 fps. Thus, it has greater power than a pocket pistol load, so the weapon should not be confused with a pocket pistol.

This weapon has some unusual qualities. First, it uses a stripper-clip system to load. In an age before separate detachable magazines dominated handgun loading, the idea of using stripper clips to load must have been attractive. In fact, it was not until World War II that rifles using readily detachable box magazines were commonly found. The handgun has problems, however, since it is difficult to load without the stripper clips. But if you have them (as anyone who carried one in the Austrian army during World War I would have, or if you finished your business within the weapon's capacity of 10 rounds), it really wouldn't matter. Once you train yourself in their use, loading with stripper clips is as fast as using a detachable magazine. It does, however, preclude a tactical reload wherein you top up your weapon during a lull in the action. And there is the problem of dropping a loaded clip in the dirt—which a box magazine avoids. But it does has a positive effect: the stripper-clip method avoids most malfunctions in autoloaders because they are primarily the result of faulty or dirty magazines.

On the formal range, you had to pull up the trigger until you felt it click, then you concentrated on the last part of the pull until the weapon fired. The barleycorn front sight is very small, as is the rear. Formal range results were not overwhelming, but this pistol was not designed for target shooting but, rather, to be used as a fighting handgun by men. On the cinema range, the trigger system came into its own. Rapid repeat shots were possible, and felt like a typical double-action revolver rather than a self-loader. The trigger system is very similar to that found in the current Austrian service pistol, the Glock 17, except that on the Glock, once the first part of the trigger is pulled and unless the trigger is fully released, subsequent shots require only that the shooter pull the last part of the trigger. On the Roth-Steyr M1907, the trigger must be pulled firmly through the whole cycle each time, as with the double-action revolver system. In this way, it is much like the DAO Beretta, Smith & Wesson, and SIG autoloaders only much smoother. It is slightly slower than with a Glock 17, but no slower than a double-action revolver, which in good hands can be fired, as Ed McGivern established, faster than self-loaders.

The beauty of the system is that after you insert the stripper clip, push down the contents with your thumb, and pull it out, the bolt will close. There are no safeties to worry about. In a time when safeties were either too difficult to disengage (thus reducing your ability to rapidly respond) or were basically unsafe, this shows remarkably good thinking. Mounted soldiers could fire this weapon, and as long as they kept their fingers off the trigger, did not have to worry about its accidentally firing if dropped. You also do not have to worry about engaging a safety. All you need do is remove your finger from the trigger to render it safe. And to get it to fire, you have to pull it completely through a firing cycle, which means it is unlikely to fire accidentally merely because you rested your finger on the trigger, as can happen with conventional single-action trigger designs.

Recoil is quite light, much like a .30 Luger. Grip angle is good and assisted rapid indexing and instinctive firing. Accordingly, I obtained excellent results on the cinema range. Because it was on a long cylindrical barrel, the front sight permits good indexing on target, although a dab of white paint would improve things a bit.

All in all, this is quite an impressive pistol. In a better caliber and with a rapidly detachable magazine to permit tactical reloading (assuming you had good, clean magazines), this pistol has a lot to recommend itself as a serious fighting handgun. Certainly, the 8mm Steyr cartridge cannot be depended upon with conventional ammunition to be a fight stopper (although the thought of THV 8mm Steyr ammunition is intriguing), but in its day this was a good fighting handgun. People who denigrate this model either have never fired it under combat conditions or really do not understand what is needed for a fighting handgun. This is one of the most surprising pistols I tested; it seems heavy, bulky, and awkward until it is actually used where it belongs: in combat situations. It then shows it is an excellent example of the breed.

I like this pistol and will keep it. If this was available in a more commonly available caliber, such as the 9mm Parabellum, I would shoot it more frequently. With good ammunition, I would not be reluctant to carry it as a duty pistol even today. Of course in 9mm, except for the machining and detachable magazine, it is a Glock 17 in many respects, and the Glock 17 is, in my considered opinion, the best 9mm pistol around today for a military handgun.

World War I Austrian soldier armed with the Roth-Steyr.

World War I Austrian soldier carrying a M1907 Roth-Steyr (above).

The author testing the M1907 Roth-Steyr 8mm pistol (right).

Left side of M1907.

The 50-foot test target shot with M1907.

The author loading M1907 Roth-Steyr with issue stripper clips.

Steyer "Hahn" M1911/M1912

9x22.7mm
(9mm Steyr, 9mm Bergmann-Bayard)

This pistol was adopted by the Austro-Hungarian government in 1912. Hahn refers to the hammer, distinguishing it from the M1907 Steyr. It is chambered for the 9mm Steyr cartridge, which is nearly 23mm long, and thus it is a somewhat larger cartridge than the more familiar 9x19mm. The Steyr Hahn was also exported to Chile, and many examples of ex-Chilean military weapons are found in the United States. During World War II, the Germans rebarreled many M1912 Steyrs for 9x19mm and these pistols are typically marked "O8" on the receiver. It is a pistol that was widely distributed during the interwar period and, of course, could have gone everywhere German troops were found, although it was primarily issued to German troops in Austria.

It was designed in that era when the detachable box magazine had not been universally accepted, and like the Roth-Steyr 07 and Mauser

SPECIFICATIONS

Name: Steyr "Hahn" M1911/M1912

Caliber: 9mm Steyr

Weight: 2.1 lbs.

Length: 8.5 in.

Feed: 8-round

Operation: NA

Sights: Blade with notch

Muzzle velocity: 1,112 fps

Manufacturer: Steyr

Status: Obsolete

M96 used stripper clips to load the magazine. This, of course, limits the user's ability to top off a magazine during a lull in combat but did avoid the problem of damaged magazine lips. The pistol uses a rotating barrel lock instead of a Browning-style lock and seems to be more than adequate for the powerful 9mm Steyr load. I imagine it takes more careful machining to make this lock than it does on a Browning-designed system. The pistol is now available in semiautomatic only, although a burst-fire version with a longer, 16-shot magazine is known to have been issued for air crews in the early stages of the war before machine guns were mounted on the airplanes.

The front sight is typical barleycorn, small, and dark. The rear sight is a shallow V-notch, making holding the proper elevation difficult. On the cinema range, the front sight was difficult to pick up rapidly, although the shiny metal rear sight block did aid triangulation and indexing.

The trigger was quite abrupt on the gun tested, and it takes care to avoid unintentional shots. The trigger is a single-action type only, and there is a side-mounted safety on the left. The safety is smaller than that found on the Colt Government Model but better than many other safeties of the period, such as the Browning '03 model.

However, this pistol has a serious design flaw that could prove quite dangerous: when you push the safety off and pull the trigger, the weapon fires. Then, when you fire a few more shots, your thumb tends to bump the safety upward just a little—not enough to engage it but enough so that the weapon will not fire. If you then put your thumb on the safety and push it down again, without touching the trigger at all, the weapon will fire. Obviously, this could prove costly. Apparently, engaging the safety only slightly will cause it to drop the hammer when pushed down firmly a second time. When this happened to me the first couple of times, I thought I was imagining it. Finally, convinced that this indeed was happening, I assumed it was a problem peculiar to my pistol and took it to my gunsmith for repair. He disassembled it and found nothing amiss. About that time, I read of another writer's test with a Steyr Hahn describing same thing. From this, I concluded it is an unacceptable design flaw.

Although I like the 9mm Steyr caliber better than 8mm Steyr, this is by no means the combat pistol that the Roth-Steyr 07 is. The best thing about these guns, besides the caliber and workmanship, is that the last ones were made in 1918, and the ownership trail has long grown cold. Use this weapon only if you cannot get anything better, watch your muzzle at all times, and remember the design flaw.

The author test shooting the Steyr Hahn M1911/12 9x22.7mm. Note the thumb position. It caused the safety to engage and a dangerous design flaw to be discovered (above).

Right side of Steyr Hahn M1911/12 (left).

Glock 17

9x19mm
(9mm Para)

The Glock 17 was first adopted by the Austrian army, then by the Swedish and Norwegian armies, and now has spread throughout the world. The adoption of the Glock 17 came despite the media hoopla about its being a plastic gun able to pass through metal detectors, which is, of course, ridiculous. But the Glock 17 is a lot more than just a synthetic-frame pistol.

The Glock 17 is probably the best 9mm *military* handgun on the market today. On the formal target range, the sights were quite big for target shooting, and the trigger pull is quite gritty. This trigger system uses a self-cocking mechanism very similar to the Roth-Steyr M907: you pull up on the trigger partway, the weapon cocks itself, and then you continue the trigger pull to fire. But it is not like a conventional double-action trigger, in that it is lighter and fires more readily. Yet it is accomplished without any shifting of the trigger, thus avoiding the typical double-action self-loader problem. The Glock offers you a degree of safety not typically found on single-action autos because you do have to pull up the double stage before firing. It is a quite rapid, yet safe trigger system. The pistol has no conventional safety; thus you need not worry about forgetting to flip it off. But it still will not fire unless the trigger is pulled through both cocking and firing functions.

Although the trigger pull is not ideal for formal target work (although not bad by any means) and the sights were quite big on the cinema range, these problems don't interfere with its combat effectiveness. You do not even notice the trigger pull for combat work. The sights are quite good for rapid indexing; they are white and allow good use in the dark. It is a quick-shooting pistol, and the pistol did not seem to recoil as much as many other 9mm ones,, although the Glock 17 weighs probably two-thirds of what other 9mm pistols weigh. Because of the Glock's

SPECIFICATIONS

Name: Glock 17

Caliber: 9mm Parabellum

Weight: 1.4 lbs.

Length: 7.21 in.

Feed: 17-round box mag.

Operation: Short recoil; self-loading

Sights: Fixed; front blade, rear notch

Muzzle velocity: 1,138 fps

Manufacturer: Glock GmbH

Status: Current production

sturdy plastic frame, you get the benefit of having a double-column magazine without the drawback of a wide butt, as is common with the Browning High Power, SIG P226, and other handguns. The Glock, despite having a 17-round magazine, is not any wider than many other conventional single-column 9mm pistols. This plastic frame that allows such light weight also absorbs much of the apparent recoil.

Possibly, the only thing wrong with the original Glock 17 was that it was available in 9x19mm only. However, it is now available in .40 Smith & Wesson, 10mm, and .45 ACP.

Accuracy was not as good as that of a P7 or the SIG P210; groups ran 4 inches at 50 feet, but this is sufficient for combat purposes. The impaired accuracy may be due to the fact that the lock up is not as firm as with either a fixed-barrel system, such as a P7, or a carefully hand-fitted barrel, such as the P210.

Because of its bulk, it is possibly not an ideal battle pistol for the infantry officer who carries a rifle, but for others armed with handguns only, such as helicopter pilots and artillerymen, it is perfect. Its weight allows them to carry the weapon with little burden, particularly with the right holster.

All in all, this Glock 17 has to be rated the number-one military handgun today. It is rapidly stripped, efficient to manufacture, low in cost, holds 17 rounds, has light recoil, and has an ideal safety system. The more you shoot it, the more you like it. I have probably put about 16,000 rounds through my own Glock and am quite pleased with it. This is clearly one of the waves of the future as far as handgun designs are concerned. I recommend it highly for military applications, although I do not think it as good for civilian law enforcement tasks.

The author range-testing the Glock 17.

Right side of Glock 17 9x19mm.

Left side of Glock
17 9x19mm.

The 50-foot test target shot
with Glock 17 (left).

Glock 18

9x19mm

(9mm Para)

The Glock 18 was reportedly designed at the request of the Austrian antiterrorist unit "Cobra." There is a simple modification to the Glock 17 to permit fully automatic fire. Unlike many such machine pistols, it does not have a three-shot burst limit, but, rather, will fire until the magazine is exhausted if the trigger is depressed. A special 33-round magazine was designed for use with this pistol, although it will also fit the standard Glocks. Unlike many fully automatic modifications of standard service pistols, the Glock 18 was not modified to take a shoulder stock: it is fired the same way as a Glock 17.

In tests I found the rate of fire to be about 1,200 rpm, although this varies according to the ammunition used. The synthetic frame of the Glock seems to soak up the recoil, and it is not any different in full auto. In side-by-side comparisons between the Glock 18 and the fully

SPECIFICATIONS

Name: Glock 18

Caliber: 9mm Parabellum

Weight: 1.4 lbs.

Length: 7.21 in.

Feed: 17-round

Operation: Recoil; select fire

Sights: Front post type; rear fixed blade

Muzzle velocity: NA

Manufacturer: Glock GmbH

Status: Current production

automatic P35, I found it the Glock much easier to control. I tested about 1,000 rounds and found reliability equal to that of the Glock 17—that is to say perfect. I have no reason to believe that it would be any different over an extended period of use.

With machine pistols, the question always arises as to what they are good for, because they are hard to control. Obviously, such weapons should be used only by well-trained, highly motivated individuals who practice a lot. As an aside, in two places where I know they are used—an island in the Caribbean by the police force who previously had Webley .38s, and in Italy by the airport police—I have some doubts about the ability of those who carry them. However, for some purposes, the machine pistol is the weapon of choice. For the first man on a raid team who will be going through a door—especially if he is carrying a body bunker, as is common in police raids—a pair of Glock 18 pistols, one in the hand, the other in a right-hip holster, would be ideal. He has a high volume of fire available, yet the ranges are likely to be very short. I found that at 5 yards, I had no problem keeping an extra 33-round burst in the one area of a "terrorist" target not shielded by a "hostage," his shoulder.

For the military, the Glock 18 would be useful for prisoner snatches or an HQ raid. It also could be useful for downed airmen because they would have the benefit at short range or at night of having a fully automatic weapon, while avoiding the usual weight of such equipment. The fully automatic option does not have to be used unless called for, but, if needed, it is there. It is noteworthy that those who are properly trained on the Glock 18 keep them on semi and switch to full auto only when it is beneficial; thus, they don't have the drawbacks from a selective-fire gun but do get greater tactical flexibility. Whereas those who have no business having such weapons tend to keep them on full auto and hope to make up in noise and bullets what they lack in skill.

The Glock 18 is probably not an ideal general-issue military choice, but for those who have the skill and tactical requirements nothing surpasses a Glock 18.

The author firing Glock 18 on full auto at hostage target at 5 yards. Note the four cases in the air.

Note that no bullets hit the target when the author fired at full auto from 5 yards.

Right side of Glock 18
9x19mm machine pistol.

Left side of Glock 18
with the selector in
full-auto and
semiauto positions
(above and left).

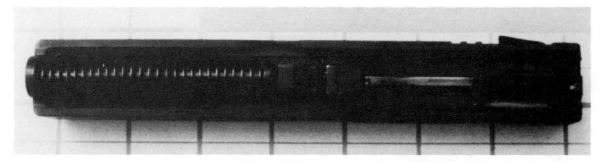

The underside of Glock 18 showing how the selector functions.

Glock 18 fitted with 33-shot extension magazine.

Browning M1910/22

7.62x17mmSR
(7.65mm Browning, .32 ACP, .380 ACP)

The Browning M1910 was designed to be a pocket pistol, and although it did find some favor with individuals who sought wartime protection, it was not until the M1910/22 version came out that the model got military acceptance. Basically the models are identical except that the barrel and butt are both lengthened on the M1910/22, and it was available in 9x17mm (.380) as well as 7.62x17mmSR (.32 ACP) caliber. As such, it was obviously more suitable for military purposes. This pistol was adopted by the Dutch and Yugoslav pre–World War I armies, as well as being used in other countries to a limited degree. More than 1 million M1910 pistols and a large quantity of M1910/22s were made, and a slightly modified model was also manufactured in Spain, so you are likely to encounter this weapon anywhere the world.

The front sight is small, shallow, and thin, and the rear sight is also quite small. These features

SPECIFICATIONS

Name: Browning M1910/22

Caliber: 7.65mm (.32 ACP; .380 ACP)

Weight: 1.3 lbs.

Length: 6 in.

Feed: 7-round, detachable mag.

Operation: Blowback, semiauto

Sights: Front blade; rear notch

Muzzle velocity: 972 fps

Manufacturer: Fabrique Nationale Herstal SA

Status: Out of production; occasional use

make it difficult to properly index the weapon on the cinema range. The safety is well placed and with practice it could be rapidly flipped on and off, but I would not trust a cocked-and-locked carry on a striker-fired weapon such as this. As might be expected on a steel-framed .32 ACP pistol, recoil was quite low, but a surprising amount of flash was evident. It is critical in all combat handguns to select a low-flash loading. This is frequently overlooked, but a properly specified load should include low-flash characteristics. Recoil in .380 was heavier but, again, nothing to be concerned about. This pistol weighs almost as much as a Glock 17 or Colt Commander .45, making the weight-to-power ratio bad.

The pistol lacks a hold-open device to indicate when the pistol is empty. I realize that with a nine-shot capacity in the M1910/22 and the use of tactical reloading, this may not be critical, but it can be easily designed into a pistol and its absence seems difficult to justify.

This pistol would seem to be a poor military choice, given its sights, caliber, safety, and weight-to-power ratio. However, we must recall that in 1922 Germany was effectively out of the weapons business, Britain made Webley revolvers (which, while excellent, were unlikely to be acceptable to the Continental militaries), and the only readily available semiautomatic pistol in production was the Colt Government Model. The M1910/22 was better than the M1915 Beretta or "Brixia" pistols from Italy, and a standard .380 was really almost as powerful. Perhaps the mistake these European armies made was not waiting for something better, but they probably figured they could always upgrade. Of course, they couldn't foresee the Depression in the United States cutting off funding, the rise of the submachine gun reducing the importance of the handgun, and higher-priority military projects like tanks, artillery, and airplanes forcing the armies to use these weapons long after they were obsolete.

Not of an outstanding design by any means, these weapons are somewhat lacking in power and are less than totally user friendly. But they may be found almost anywhere, and the Belgian examples are well made. They also have a smooth, slick appearance and are sufficiently accurate for practical purposes. Carried in condition three, they are safe, if slow, and the concealment may make up for their lack of speed.

The author firing M1910/22 7.65mm.

Browning M1910/22 with 50-foot test target.

Browning P-35
9x19mm
(9mm Para)

John Browning designed this pistol, with the finishing touches put on after his death. It was considered by many to be a culmination of Browning's career.

The Belgian army adopted it prior to World War II to replace the M1903 9x20mmSR long pistols. For many years, it was the only double-column 9mm handgun in production, thus giving it the highest capacity of any pistol being made. I well remember the years when people used the Browning Hi-Power because they liked the fact that it would hold 14 rounds, so that they were basically ready for "war." Many thought if you carried 14 rounds in your pistol, you did not need to worry about spare ammunition or reloading, and there is something to be said for that, I suppose. However, a steel-frame pistol like the Browning necessitates a fairly wide grip, and, as a consequence, the trigger system has to be

SPECIFICATIONS

Name: Browning P-35

Caliber: 9mm Parabellum

Weight: 2 lbs.

Length: 7 3/4 in.

Feed: Double-column box

Operation: Recoil

Sights: Blade/notch

Muzzle velocity: N/A

Manufacturer: Fabrique Nationale Herstal SA or John Inglis Co. Also made in Argentina, Indonesia, Israel, and Hungary.

Status: Current production; worldwide use

set up differently than on the Colt Government Model, thus producing problems with the trigger pull.

The example that I tested was a Canadian-manufactured Inglis with accompanying wooden stock holster. This particular stock was manufactured in Canada under Chinese army contract to be sold during World War II through the U.S. Lend-Lease program. However, the standard Browning is common throughout the world with a flat, broad stock. The grips are too thick for comfort, and the sights are very poor. Despite an adjustable target rear sight, the rear notch was much too narrow and difficult to use. On standard nonadjustable sights, they are also quite narrow and difficult to see. Likewise, the front sights are too low, too small, too narrow, and difficult to use properly on both the formal target range and the cinema range, where the shallowness and smallness of the sights make indexing difficult. I was able to get 3 1/2-inch groups.

This pistol has a shape more like a conventional pistol, and thus the addition of a butt stock really does not make that much difference in shooting accuracy, unlike the Broomhandle Mauser Military Pistol where the use of a stock reduces the group almost by half. Four of the five shots, however, did go into 2 inches, so the pistol is capable of pretty good work. The sights are just so difficult and the trigger so hard to use that it is difficult to bring it out. At extended range, I found that about 100 yards was the accuracy limit, with or without the butt stock. I also found that with the factory 9mm ammunition I used the adjustments on the sights were off by almost 200 yards. That is, when I put it on 300 yards, I was shooting at 200 yards and hitting, and when I had it on 400 yards, I was shooting at the 300-yard range.

The P-35 Hi-Power is common throughout the world: the German army used it during World War II, and tens of thousands of them have been manufactured since. I have no doubt that these pistols are reliable because they use a typical Browning-type design. However, it does have some shortcomings. The safety on all the factory guns is much too flat and difficult to push off, making it hard to disengage with any rapidity. To compensate, most people either carry the weapon with the hammer down and a round in the chamber and then try to cock the hammer (which is always slow and inefficient) or they fit larger nonstandard safeties, which allow a bigger platform from which to push off. On military weapons, however, you have to take them the way you get them, flat safety and all.

Although it is reliable, the P-35 is heavy for its caliber, the grips are thick, the safety is somewhat marginal for military purposes, and you have the same problem that you have with all single-action autoloaders. Most examples have magazine disconnectors, which I think is probably a good, useful feature. Many of them have drop-magazine features that prevent the magazines from falling out completely, and to get the empty magazine to fall out properly you have to bend the spring out of the way. It is not nearly as good as some weapons that are not as well thought of (e.g., the M50 French 9mm). Given a

The author testing Inglis-made Hi-Power without stock.

The author firing Inglis-made Hi-Power with Chinese army-issue butt stock affixed.

Right side of Inglis-made Hi-Power as supplied to the Chinese during World War II.

Inglis-made Hi-Power 9x19mm group shot at 50 feet without stock.

Groups shot at 50 feet with stock.

choice between the French Mac 50 and the P-35, both stock, one would probably be better off with the French M50 because the safety is much easier to disengage. Certainly the Hi-Power has a slightly greater capacity but for most of the situations, that capacity is unnecessary and results in awkward grip and a poor trigger system.

Still, despite its design shortcomings, people just love these pistols. My friend Leroy Thompson, internationally known in counter-terrorism circles, is a strong supporter of the Browning Hi-Power. The FBI "Super SWAT" National Hostage Response Team (HRT) carries Browning Hi-Powers, as do the British Special Air Service (SAS) units. If you are offered the choice, however, you could probably do better than a Browning Hi-Power. I prefer an Astra 400 for a serious combat gun. The safety is easier to get off, and the weapon can be used in a combat situation more effectively than the P-35.

Colt New Service

11.43x32.1mmR
(.45 Long Colt)

This weapon in .45 Colt was standard issue in the Canadian navy during World Wars I and II, and British forces used a version chambered in .455 Webley or .455 Colt. A slight modification to the cylinder allowed .45 ACP cartridges to be used, and U.S. forces substituted this as its standard pistol during both world wars.

On the formal target range, the sights are hard to see because of their thin front blade and small rear notch. Still, I managed to fire a group slightly smaller than that fired with our Smith & Wesson control gun at 50 feet, showing what the weapon can do if the shooter holds it correctly.

The Colt New Service has always been known as a "man's gun": it is chambered in a wide variety of man-stopping calibers, it is large, and the parts are rugged and durable. It has always enjoyed a reputation as a long-lasting piece. Even today, with pistols that are frequently 90 years old, they soldier on as good as the day they were made.

This same rugged frame makes the pistol grip too large for my size-9 hands. On the cinema range, I had to shift the pistol in my hand to pull the trigger, which pulled the weapon to the side and caused the shot to go wide. The high, thin front sight allows good indexing in poor light, although it would be better if painted white. The rear sight is too small and shallow, thus reducing one's ability to triangulate quickly.

I found that reloading the Colt is slow because the cylinder release must be pulled to the rear and then flipped out, whereas on the Smith & Wesson, one continuous push gets the job done more quickly. Recoil was quite heavy in this .45 Colt chambered weapon, and this slowed repeat shots. I found that I could fire three shots with the Glock 17 in the same time it took me to fire one shot and recover with the Colt New Service. The thin grips at the top caused

SPECIFICATIONS

Name: Colt New Service

Caliber: .45 Long Colt

Weight: 2.4 lbs.

Length: 9 3/4 in. with 4-1/2 in. barrel

Feed: Revolver

Operation: DA

Sights: Blade/notch

Muzzle velocity: N/A

Manufacturer: Colt

Status: Obsolete

my thumb joint to be pounded on each shot, which made prolonged strings unpleasant. The ejector rod was too short to fully eject the long .45 Colt cases, although obviously this would not be a problem with the shorter .455 Webley or .45 ACP cases.

The weapon is quite large and heavy, probably too heavy for an infantry weapon in the 1990s, but for World War I's non-rifle-carrying officers, it was an acceptable pistol, especially for officers with large hands. It was safe to handle and rugged enough to hold up well. Herbert McBride tells of a friend who, during the war, took a pair of Colt New Service revolvers with him and did good work with them. It is hard to disagree with such success.

These rugged weapons are likely to be found in the world wherever men of adventure have been. Even with the finish gone, the weapon will perform, and all of the calibers are good ones.

Right side of the 5 1/2-inch Colt New Service .45 revolver.

CZ 24

9x17mm

(.380 ACP)

This pistol preceded the more common and popular CZ 27 7.65mm model and suffers from all the same handicaps. But, unlike the CZ 27, the CZ 24 is available in .380 caliber. The Czechs always seemed to like more powerful weapons than their Central European cousins. Even if they used the same caliber, they loaded it hotter than anyone else; witness the .30 Mauser and 9x19mm Czech loadings.

The front sight on the CZ 24 is European barleycorn with V-shape rear hatch. It is very hard to hold elevation, and groups were only of average size. Trigger pull is good because the parts have been carefully fitted. Both from the height and width standpoint, the small front sight makes indexing on the cinema range difficult. Additionally, the narrow V-shaped rear sight creates its own difficulties on the cinema range: you do not get enough light (as seen through the rear sight) around the front blade to make indexing easy.

SPECIFICATIONS

Name: CZ 24

Caliber: .380 ACP

Weight: 1.5 lbs.

Length: 6 in.

Feed: 8-round, in line

Operation: Recoil; semiauto

Sights: Front blade; rear V-notch

Muzzle velocity: 984 fps

Manufacturer: Ceska Zbrojovka, Brno

Status: Obsolete

The safety is hard to disengage with the thumb because of thick grips. On the cinema range, this difficulty caused me to miss one shot that would have resulted in my death in the real world. Although the safety is difficult to disengage, cocking the pistol is also difficult because of the shape of the grip and strength of spring. Recoil is low, as is to be expected in a steel-frame .380, but, pleasantly, no hammer bite was experienced.

The Czechs soon realized that this design was not suitable for a military handgun and went to the self-cocking CZ 38 .380. Had they made the CZ 38 into a 9x19mm pistol, it would have been a world beater, much like the Glock 17 we are all familiar with.

The design of the CZ 24 represents a dead-end for a military handgun. About the best that can be said for it is that it was no worse than most of the .32 ACP self-loaders common during and just after World War I—but at least it was a .380, which has to make it better if all other factors are equal.

The author field-testing the CZ 24. Note the case-clearing ejection port and recoil.

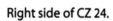

Right side of CZ 24.

Left side of CZ 24.

The results of the CZ 24's
50-foot target shooting.

CZ 27

7.62x17mmSR
(7.65mm Browning, .32 ACP)

This pistol is a perfect example of how weapons or systems that feel good on the target range may not be so good on the cinema range. This pistol is heavy with a steel frame and steel slide. The pistol I tested was in mint condition, of wartime production, and Waffenamt stamped.

The serrations on the slide are very sharp, so sharp in parts you almost cut your hand. However, the hammer is well shrouded by the slide and does not catch on clothing. The butt magazine release closes the slide when the magazine is withdrawn and, of course, delays the withdrawal of the magazine because of the pressure on the magazine from the slide. This also means that if you remove the magazine, you have to pull the slide to the rear to reload. The heavy spring tension on the slide, coupled with the serrations, makes it this difficult.

The sights are poor and the trigger is very diffi-

SPECIFICATIONS

Name: CZ 27

Caliber: 7.65mm Browning (.32 ACP)

Weight: 1.6 lbs.

Length: 6.3 in.

Feed: 8-rd., in-line, detachable mag.

Operation: Blowback; semiauto

Sights: Front blade; rear V-notch

Muzzle velocity: 919 fps

Manufacturer: Ceska Zbrojovka, Brno

Status: Obsolete

cult, decreasing accuracy. My shots went into a 5 7/8-inch group at 50 feet, which is roughly twice as big as the Smith & Wesson Model 19 revolver target used as my reference.

The safety feature on this particular pistol is quite interesting. The button on the left-hand side of the weapon that appears to be a magazine release is actually a safety release. You push the lever down to engage the safety; then to disengage the safety, you push the button and the lever flies back up. On the formal target range where you have plenty of time and light, this is accomplished quite easily: you put your thumb on the button and push inward with your knuckle and disengage the safety. However, on the cinema range with poor light conditions, disengaging the safety is more difficult because the button itself is quite small and you have to must press directly on the safety, not at an angle, to disengage it.

This safety feature creates another major

The author field-testing the CZ 27.

Left side of CZ 27 7.65mm.

Pushing in with the knuckle of the thumb, the shooter can disengage the CZ 27's safety without shifting the weapon in his hand. Once mastered, this technique is quick, but trimming the grips would make it faster and more certain.

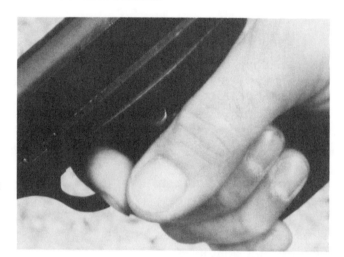

The CZ 27 50-foot test target.

concern: how to carry this pistol. You have three basic options. First, you can carry it with the hammer down, chamber empty. But if you do it that way, the pistol is obviously much too slow to get into operation, and the serrations on the slide cut your hand more than on a Government model. Second, you can carry it with the chamber loaded and the hammer down. But you have a lot of tension on that spring because of the highly shrouded hammer and the coil hammer spring, very little surface area to grab, and a spur-type hammer, making it difficult to cock. Third, you can carry it cocked and locked. But this is very dangerous because even a slight pressure on the safety button will cause it to flip off. I would not trust a weapon

cocked and locked like this in my pocket. You would need a specially designed holster to carry the pistol in this manner—designed by someone familiar with the problem, such as Ken Null, who would cut out the area around the button and configure the holster so that when you retrieve the weapon the safety button is not inadvertently pressed by the leather. Needless to say, there aren't too many Null-made holsters around for a CZ 27.

The CZ 27 is made from good materials, but the caliber is too light for conventional military ammunition. The weapon itself is heavy and slow to get into operation, and it has an ergonomically problematic safety, a difficult trigger, and poor sights. All in all, this not a good handgun for military purposes.

CZ 38

9x17mm
(.380 ACP)

This is one of those pistols that is better in practical use than it is on a formal target range. The CZ 38 is a double-action-only pistol (actually self-cocking), making target groups very difficult to fire. I achieved 3 3/4-inch groups with this weapon using modern full-jacketed ammunition.

Aside from the caliber, however, this is a great military weapon. The self-cocking-only trigger makes it safe. It is a flat weapon, and with no projection to catch on clothing and no safety to worry about; it will not go off unless you pull the trigger. All this adds up to a good combat gun under stress.

The weapon does have a few undesirable features. The slide closes when you pull the magazine out, but it functioned perfectly in all firing tests. The sights are difficult to see on the cinema

SPECIFICATIONS

Name: CZ 38

Caliber: 9mm (.380 ACP)

Weight: 2 lbs.

Length: 6.3 in.

Feed: 8-rd., in-line, detachable box mag.

Operation: Blowback; semiauto

Sights: Front blade; rear V-notch

Muzzle velocity: 1,000 fps

Manufacturer: Ceska Zbrojovka, Brno

Status: Obsolete

range and thus difficult to index, but painting them white takes care of this problem.

The pistol I evaluated was manufactured for the German army during World War II.

The CZ 38 was originally developed to Czech air force specifications for a defensive weapon, thus its clean, snag-free lines.

It is a superior .380 caliber pistol, somewhat large for the caliber, but easily stripped and put into operation. I would rate it as my favorite of the .380s. I liked the self-cocking-only feature, the convenient oversized trigger guard that is easy to use with your gloved hand, the easy stripping, and the fact that it is flat and safe to use without any safety worries. With good .380 ammo, it is a much better pistol for military use than the common Walther series of pistols.

The author test-firing a CZ 38 .380 pistol.

The hammer on the CZ 38 is an excellent design feature that avoids snagging on clothing.

The rear view of CZ 38 .380 shows its thin slide and receiver, which aids in concealed carry.

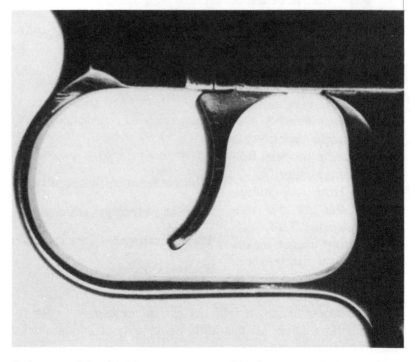

A close-up of the CZ 38's trigger system. I like the oversized trigger guard.

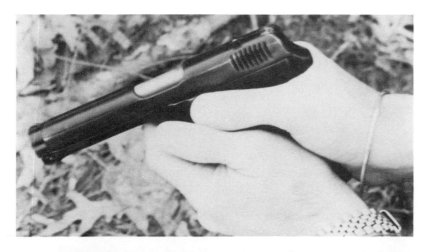

In this photo, you can see how the hammer spur prevents "biting."

The CZ 38 strips easily and rapidly to expedite cleaning.

The 50-foot test target shot with the CZ 38.

CZ 45

6.35mm

(.25 ACP)

Although this pocket pistol might, at first glance, seem out of place in this volume, Czech and Soviet intelligence agents commonly used it during the Cold War. My friend Leroy Thompson can verify this; he once took one off a Czech defector he was assigned to guard. The CZ 45 is actually very similar to the CZ 36, which was used by Czech pilots during World War II, so I think this pistol deserves mention. Additionally like all Czech pistols, it is (or has been) sold widely over the world, and thus you are likely to encounter it in various out-of-the-way places. Plus, it is unlikely that the pistols you encounter will have much of a documentable history, always helpful.

This all-steel pistol reminds me very much of the Seecamp double-action .25/.32 ACP self-loader. It has no safety, fires in the trigger-cocking mode only, and has a magazine safety. Like the

---

SPECIFICATIONS

Name: CZ 45

Caliber: .25 ACP

Weight: 13 3/4 oz.

Length: 4 4/5 in.

Feed: In-column, single-feed box

Operation: Blowback

Sights: None

Muzzle velocity: NA

Manufacturer: Ceska Zbrojovka, Brno

Status: Limited issue

Seecamp, it has no sights and thus does snag on your pockets. This lack of sights means it is difficult to shoot on the formal range, and it was only a desire to have the results consistent that caused me to fire it at 50 feet rather than the 10 feet that would be more common for such a weapon. Although the group shot at 50 feet was at least twice the average group shot with combat handguns, the fact that I was able to hit it at all much less the good (relatively speaking) group I got speaks well of the design of the weapon.

The shape of the spur on the grip keeps the slide from slicing your hand, as happens(at least to me) with Walther TPH/PPK-style pistols. The trigger pull is heavy but smooth, which facilitates rapid repeat shots. The heel butt magazine release helps keep the width small and avoids accidental release when carried in the pocket.

Oddly enough, on the cinema range, the lack of sights became much less critical. The groove in the slide created a shadow effect much like that found on the Heckler & Koch VP 70 pistols. I was able to align my pistol quickly and follow up with the sights. I would not have realized this feature had I not shot it on the darkened cinema range.

The CZ 45 is obviously a very fine pocket pistol, but I have some great doubts about its military efficiency. The caliber is simply too light. As a badge of office or a killing tool issued to intelligence agents who shoot their enemies at close range in the head or to others who work in urban areas where weapon noise is a consideration, it is sufficient. The double-action-only trigger system makes for a safe weapon in the hands of nervous and less-well-trained shooters. Frequently, intelligence agents are armed but not really "gun people." As a spare weapon or something to back up a battle rifle, this pistol would be fine. As primary armament, I think it is too light. It is only fair to note, however, that the most successful German pilot in World War II, who operated on the Russian front, carried a .25 ACP pistol in case he got shot down, so perhaps I am thinking too much like an American who likes a .45!

The author test-firing the CZ 45 .25 ACP.

The right side of the CZ 45.

Left side of CZ 45.

A lack of sights resulted in 50-foot target groups double the average size.

CZ 50

7.62x17mmSR
(7.65mm Browning, .32 ACP)

The CZ 50 is a pocket pistol designed and manufactured in Czechoslovakia during the post-World War II era. It is commonly seen in the hands of military and police organizations in Czechoslovakia and has been exported in considerable numbers. In Czechoslovakia, military officers were not allowed to carry the Vz 52, the Czech service weapon chambered for the 7.62mm cartridge, when not in uniform. Officers who wished to carry an off-duty gun typically opted for a .32-caliber autoloader, and the CZ 50 was a common choice.

The CZ 50 is a selectable-trigger-action autoloader pistol. One can cock the action, put the safety on, and carry it cocked and locked or, alternatively, drop the hammer and use it like a double-action revolver. That certainly has some advantages.

These weapons were not common in the

SPECIFICATIONS

Name: CZ 50

Caliber: 7.65mm Browning (.32 ACP)

Weight: 1.5 lbs.

Length: 6.8 in.

Feed: 8-rd., detachable box mag.

Operation: Blowback; semiauto; DA only

Sights: Front blade; rear notch

Muzzle velocity: 910 fps

Manufacturer: Ceska Zbrojovka, Uhersky Brod

Status: Current production; many exported

United States, because of the prohibitive import duty, until the recent "Velvet Revolution." In places like South Africa and Latin America, however, they are common.

I read a 1950s report that stated that this pistol was made of poor materials and had a poor finish. Possibly the one tested for that report did suffer from those drawbacks. Certainly, the one I tested did not. It was well made and well finished. I can only conclude that either that other example was of much poorer quality than mine (which I doubt) or that he was prejudiced against Czech Communist weapons (which I am not).

The CZ 50 did have some minuses. The example I tested had a couple of malfunctions that I believe were the result of a faulty magazine. The front sight is low, dark, and hard to see; the rear sight is small, which makes indexing quite hard on the cinema range. Also, the pistol, by virtue of the angle of the grip, pointed low in my hand. It is a heavy pistol for the caliber and quite bulky for the 7.65mm cartridge. There is nothing really noteworthy about this pistol except the safety. It is reversed from the typical Colt Government Model practice and is somewhat slower to use, at least in my hands, which are more familiar with the Colt Government Model systems. If you used it all the time it might not feel that slow.

It is not up to the design or quality of a Sauer Model .38 or a Walter PP/PPK, but it is certainly an adequate pistol. Group sizes for the five-shot group ran approximately 3 1/4 inches at 50 feet, which is acceptable for a small pocket pistol with the sights that it has. Perhaps if you use .32 ACP THV rounds, this pistol would be all right. Otherwise, it is marginal with military ammunition, and for that reason alone, it should be avoided if possible.

The author test-firing the CZ 50 7.65mm.

Detail of the slide and release system of the CZ 50.

The CZ 50 can be carried cocked and locked, unlike many double-action pocket pistols. This is helpful after the first shot is fired because you can avoid having to carry a loaded cocked weapon with no safety or having to drop the hammer and start all over again.

A close-up of the CZ 50's safety system.

Results of the CZ 50's 50-foot test-target shooting.

Vz 52

7.62x25mm
(7.63mm Mauser, .30 Mauser)

This was the former service pistol in Czechoslovakia. When the Soviet Union compelled its former puppet states to adopt Soviet calibers, the Czechs, instead of adopting the Tokarev, designed the Vz 52. Later, when they adopted the CZ 83 in 9mm Makarov, the Czechs surplused off quantities of these Vz 52 pistols. Consequently, they are likely to be found throughout Africa and South America and are also common in European terrorists organizations. "Carlos," the infamous terrorist, shot three French policemen in the 1980s with one of these pistols, killing one and wounding two, and he later used one to shoot a rich pro-Zionist merchant in London, severely wounding him.

These two incidents point out a serious problem with this round: the caliber lacks stopping power despite its projectile speed. It is small, lightweight, full-jacketed, and round-nosed. The .30 Mauser cartridge in the hot Czech loading,

SPECIFICATIONS

Name: Vz 52

Caliber: 7.62mm (7.63; .30 Mauser)

Weight: 2.3 lbs.

Length: 8.25 in.

Feed: 8-rd., detachable box mag.

Operation: Recoil; semiauto

Sights: Front blade; rear square notch

Muzzle velocity: 1,600 fps

Manufacturer: Ceska Zbrojovka, Uhersky Brod

Status: Obsolete; Czech army reserve use

which frequently will push the 86-grain bullets at 1,650 fps from a pistol barrel, does offer some interesting prospects for vest penetration. Even with steel-cored bullets, it is stopped in a ballistic vest, but with THV-style projectiles, better results could be anticipated. As an aside, an acquaintance of mine who served with the SAS and MI5 reports getting shot on two occasions while wearing his David Second Chance vest, once with a 9mm Makarov and once with a .30 Mauser. He said he hardly knew it when he was hit with the 9mm Makarov, while the .30 Mauser left a big bruise. On testing them, I found that normal jacketed loads went through five layers of ballistic material; steel-cored or THV-type rounds would undoubtedly do much better.

Using Fiocchi ammunition on the formal target range, I encountered a great deal of flash even in the daylight. The trigger pull was hard, and the sights were small and difficult to pick up.

The right side of the Vz 52 7.62x25mm.

Close-up of the trigger and slide release on the Vz 52.

Groups ran 2 5/8 inches. The weapon pointed extremely low, and most of my hits were in the knee area even though I aimed at the chest. Naturally, this could be overcome by training (or by lowering the front sight blade), but it is useful to note because you may have to depend on one of these pistols someday, without an opportunity to range test.

The grip is too long because of the long cartridge. Stacking them at an angle like the Type 94 Nambu would allow a shorter grip. The safety was small, stiff, and difficult to disengage. Worse, it tends to hit your thumb when in the shooting position and get bumped back into the safe, making the weapon inoperable.

I did find one interesting feature the day I tested the weapon on the outdoor range: how flat the weapon shot. After going to the formal 50-foot range, I shot at some targets at 150 yards. The .30 Mauser cartridge shot very flat in comparison to the .38 Special and .45 ACP. Holding dead-on with the .45 ACP and .38 Special, I put one shot 4 to 6 feet low. With proper ammunition, a .30 Mauser cartridge pistol could easily substitute for a Young rifle in the hands of a soldier well trained in combat pistol techniques.

On the cinema range, the bright muzzle flash slowed repeat shots by a factor of two-thirds in comparison with .38 Special weapons. The front sight again appeared too small and short for good muzzle pickup, and the rear sight was so small and dark that it was almost impossible to pick it up. Indexing with this weapon is difficult. Recoil was light because of weight of the steel frame, but the grip felt big on the cinema range.

The design does facilitate carrying the weapon in condition-one cocked-and-locked position, but the small safety requires concentration to disengage it and care *not* to bump it accidentally to the off position. Further, carrying it in condition two is unacceptable because of the difficulty in cocking the burr hammer. It forces you to completely break your grip, adding precious seconds to the time required to get the weapon into action. The weapon can be carried in condition two without having to worry about manually dropping the hammer if you place your thumb on the safety, press it firmly down. and then manually cycle the action. The hammer will follow the slide down, allowing you to get to condition two without having a "hot gun" in hand. By the way, although this pistol looks like a double-action pistol, it is not. It is single-action.

The weapon can be rapidly field stripped for cleaning, but the roller locking system requires a screwdriver to disassemble—obviously a drawback. The butt-mounted magazine release is also undesirable, especially since the lanyard is also mounted there. Once, when removing the magazine, I found I could not withdraw it because the lanyard loop jammed on the magazine release, and I could not get the magazine out until the lanyard loop was pushed clear.

Overall, the Vz 52 is not a great combat pistol, but it is as good as, if not better than, the Tokarev TT 33 7.62x23mm and without a doubt the strongest production handgun ever chambered for this caliber.

NOTE: The hot Czech loads for this pistol are *not* suitable for use in other 7.62x25mm handguns.

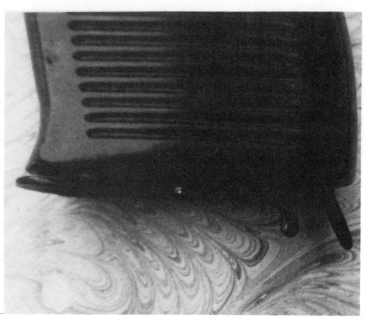

The positioning of the lanyard loop on the butt of the gun makes quick magazine release and replacement difficult.

The safety system on the Vz 52 must be carefully watched. Recoil causes the shooter's thumb to knock the safety on after the first shot.

This photo illustrates how the roller locks function on the Vz 52, causing the locking barrel to slide until pressure drops. This system is similar to those found on the MG42 and various Heckler & Koch rifles and submachine guns.

The 50-foot test target shot with the Vz 52 7.62mm.

CZ 75

9x19mm

(9mm Para)

Designed in Czechoslovakia for export, the CZ 75 is an intriguing weapon. Many experts, including Jeff Cooper, the noted .45 Government Model fan, have described it is a wonderful 9mm.

It has been adopted by numerous military organizations, as well as various counterterrorist police agencies. A friend of mine with the Paris R.A.I.D. unit carried one, and one city in England issues it as its duty weapon. It has been purchased in quantity in India, is carried by many Latin American military officers, and is common in Lebanon, Africa, and Europe. About the only place it was rare is the United States because of the prohibitive U.S. import tariffs on arms from Warsaw Pact countries, and that is now changing.

The CZ 75 is the most copied design since John Browning's pocket pistols, with direct copies being made in Italy, Switzerland, and the United States, and design spin-offs being produced all over the world.

It might be best to view this pistol as a combination of a Browning P-35 and a SIG P210. It has the double-column magazine of the P-35 but uses the slide-receiver system of the P210. It offers a double-action trigger, but unlike with most of the others, you can also carry the weapon in condition one. More to the point, you can carry it loaded with the hammer down, pull the trigger for the first shot, and then reapply the safety if you anticipate shooting again but do not want to have to restart from the self-cocking position. This is a great feature because frequently people are running around after firing a shot with a cocked auto and no safety.

The CZ 75 does have some drawbacks. The double-action trigger system makes it more complex than the P-35 single-action trigger system. Also in comparison to a SIG P210, the CZ 75 is frequently soft—not softer than many P-35s but

SPECIFICATIONS

Name: CZ 75

Caliber: 9mm Parabellum

Weight: 2.2 lbs.

Length: 8 in.

Feed: 15-rd., detachable box mag.

Operation: Recoil; semiauto

Sights: Front blade; rear square notch

Muzzle velocity: N/A

Manufacturer: Ceska Zbrojovka, Uhersky Brod

Status: Current production

just not up to SIG P210 standards (an excellent article in a recent *Handgunner* magazine details this problem). Also, because the Czech government arsenals load all of their ammunition hot, the bore is frequently a little larger than is desirable. It does not matter at first, but after about 5,000 rounds, the edge is frequently off the CZ 75, whereas the P210 just gets better. P210s are known to go 250,000 rounds without a problem.

The front sight is fine on the CZ 75, although it could be improved by painting it white. I think the rear sight is too narrow, and this hinders rapid indexing on the cinema range. The trigger pull on both double-action and single-action was good in the example I fired. As might be expected on a steel-framed 9x19mm pistol, recoil is quite light.

This pistol would rate very high on my list of combat handguns *if* it was offered with 1) an

The author test-firing the CZ 75 9x19mm.

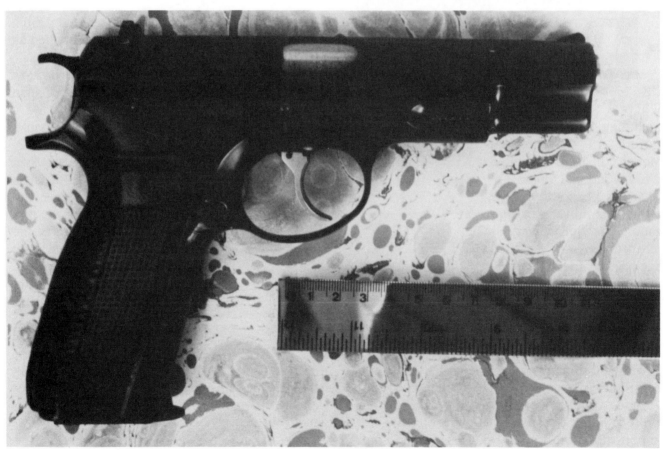

Right side of CZ 75.

alloy frame, 2) a decocker lever so the hammer could be dropped without pulling the trigger (a recipe for trouble with many military people), 3) a magazine safety, and 4) white sights. As is, the CZ 75 performs quite well, and the results are probably even better when used by well-trained people who can maximize its good qualities while minimizing its shortcomings.

The CZ 75's 50-foot test target.

CZ 83

9x17mm/9x18mm
(.380 ACP/9mm Makarov)

The CZ 83 was the standard duty weapon in the Czech army in 9mm Makarov caliber and is available for commercial sale around the world in .380. Although it is the size of many mini-9x19mm pistols available today, the typical Czech loading has about 20 percent more velocity than normal Western loadings. Therefore, you are pushing the envelope with this weapon and a Czech loading as opposed to the Western mini-9x19mm and standard low-velocity ammunition commonly encountered in the West.

On the positive side, the trigger pull is quite good, and the weapon feels good in my hand. Sights are also handy. On the cinema range, the white-colored front and rear sights allowed quick pickup in the dark and rapid indexing. Recoil was light due to the caliber and the steel frame. This feature plus the good sights allowed rapid repeat shots and follow-up.

SPECIFICATIONS	
Name:	CZ 83
Caliber:	.380 ACP (9mm Makarov)
Weight:	1.4 lbs.
Length:	6.8 in.
Feed:	NA
Operation:	Blowback
Sights:	Front blade w/ white insert; rear square notch w/ 2 white spots
Muzzle velocity:	950–1,105 fps
Manufacturer:	Ceska Zbrojovka, Uhersky Brod
Status:	Current production

One of the best features of this pistol, as well as the CZ 75, is that the shooter can put the weapon in the loaded-chamber, hammer-down mode and then fire the first shot by merely pulling the trigger. With most double-action autoloaders, once you have fired, you are faced with the prospect of lowering the hammer and starting all over from the Weaver double-action mode or having a loaded, cocked pistol in your hand, with no safety on. With the CZ 83 (and CZ 75/85) you can simply flip the safety on. Then when you are ready to fire again, flip it off and fire from the easier-to-control double-action trigger mode. If the designer would only retain that feature and then add a decocking lever like that on the SIG P220 pistol, the shooter would have the best of both worlds—he would avoid the risk of lowering the hammer manually on a live round.

One weakness I did notice was that because of the 13-shot magazine the grip was somewhat thick, forcing me to shift my grip to hit the magazine release button with my thumb. This may be of limited importance, however, because of the number of rounds the magazine holds.

I like this pistol and prefer it in 9mm Makarov to .380, especially with Czech loads, because that gives me a borderline 9x19mm equivalent. (An even better choice, however, is the Hungarian RK 59, although you do have to give up high magazine capacity to get the RK 59's lightness and handiness. I believe this is a good swap.)

Wherever good guns are appreciated and hard currency can be had, you will find Czech guns. The CZ 83 is no exception. If you are in need of a handgun and come across one in a slave market in Sudan, grab it. It is far better than some of the more popular and expensive .380s, such as the Walther PPK.

Author test-firing the CZ 83 9x17mm pistol.

The 50-foot test target of the CZ 83.

Bergmann-Bayard M1910/21

9x23mm
(9mm Bergmann-Bayard, 9mm Steyr)

This was the standard Danish service pistol through World War II. The Germans confiscated some for police use in World War II, and no doubt the underground soldiers also used some. The design was also manufactured in Spain prior to the 1930s.

The Bergmann-Bayard's cartridge is some 4 millimeters longer than the standard 9x19mm ammunition and was intended for a hotter load. The pistol is well made out of good materials, but it does reflect a time when it was not clear to everyone how the combat handgun was to be used.

On the cinema range, the 9mm Fiocchi loadings had a lot of flash showing quickly, establishing that they were not the loads for firing in the dark. The grip felt good in my hand and allowed for good indexing on the range and instinctive shooting.

The trigger is quite good on this weapon,

SPECIFICATIONS

Name: Bergmann-Bayard M1910/21

Caliber: 9mm

Weight: 2.3 lbs.

Length: 10 in.

Feed: Double-column box magazine

Operation: Recoil

Sights: Blade/notch

Muzzle velocity: NA

Manufacturer: Anciens Establissements Pieper of Herstal, Belgium

Status: Obsolete

showing how carefully the fitting had been done. The safety system is the real drawback: you must put the thumb on it, pull it down, and then take your grip back up and fire. It is a slow process, and the thumb obscures the top of the weapon, making it difficult to pick up the target. However, that system works better than trying to cock the weapon when you need to shoot, because it has a very heavy spring and a small hammer; this slows your response and forces you to break your grip doing it that way.

The forward-mounted magazine only appears awkward. By simply pushing forward, you can quickly release the magazine and replace it with a spare much faster than on the fixed-magazine models of Broomhandle Mauser/Astra weapons. Additionally, you can load the weapon with stripper clips through the top, which will save the cost of replacement magazines. It is hard for me to see why Mauser did not adopt this magazine system at the onset. Putting the magazine ahead of the trigger guard also allows you to use a longer cartridge and to have a very small grip while still shooting large cartridges. The latter is helpful to people with small hands. The method does increase the overall length of the weapon, but in a military arm this may not be a critical factor.

The wide-barrel rib facilitates indexing in poor light, although the rear of the pistol was quite cluttered, making it difficult to pick up the rear sight quickly. It is hard to triangulate, but the barleycorn front sight on a skinny

barrel did pick up well in the dim light.

This is a solid, well-made pistol. It has a nice blue finish, is well machined, and is heavily constructed in a good caliber. It would make a fine World War I choice—except for the safety disengagement—but most weapons of that period, except the Roth-Steyr 07 and Colt M1911, suffer from slow safety disengagement systems. I prefer the Mauser M96 to the Bergmann M1910/21 because the Mauser seems quicker on the cinema range for repeat shots and for indexing under dark conditions. However, I like the Bergmann magazine system better.

The author test-firing Bergmann-Bayard 1910/21.

This photos shows the magazine removed from the Bergmann-Bayard. Placement of the magazine to the front rather than in the handle allows longer and wider stacks to be used, while still retaining a comfortable grip shape.

Open action on the Bergmann-Bayard.

Top view of Bergmann-Bayard sights.

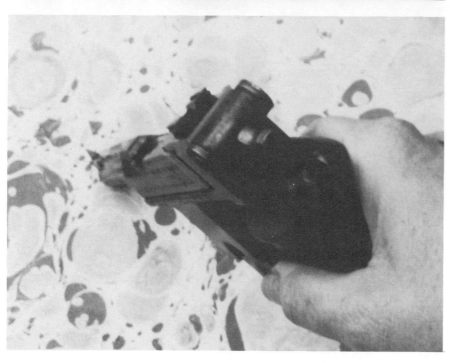

The rear sight picture of Bergmann-Bayard.

The 50-foot test target shot with the Bergmann-Bayard M1910/21.

Webley
RIC M83
11.43x19mmR
(.455 Webley Mk 2)

Although never an official military revolver, the Webley M83 is included in this survey because of the many NCOs and officers who carried the Mk2 .455 caliber handgun to war as their personal weapon, much the same as the M60 Smith & Wesson was included as a Vietnam War-era handgun.

The M83 Webley was first advertised in 1884 and continued in Webley catalogs until 1939. Various British police and colonial agencies purchased the weapon, and as a consequence, they are widespread. You are very likely to encounter a Webley M83 anywhere. Belgian copies were also widely distributed, and, except for differences in finish and quality of construction, such copies are quite similar to the British version.

Jeff Cooper once described the Webley M83 as "a great gun" because of its solid frame construction, .455 caliber, and five-shot capacity coupled

SPECIFICATIONS

Name: Webley RIC M83

Caliber: .455 Webley

Weight: NA

Length: NA

Feed: NA

Operation: NA

Sights: NA

Muzzle velocity: NA

Manufacturer: NA

Status: Obsolete

with 21-ounce weight. There are many people who would rather have a 21-ounce .455 Webley M83 than a 19-ounce M36, 2-inch .38 Special. Although the Webley M83 is bulkier than a J-frame Smith & Wesson, it is no more so than a D-frame Colt or one of the small-frame Ruger pistols. The Webley M83 lacks the rapid ejection system of the swing-out cylinder or break-top revolver, but unless you are planning on rapidly reloading your revolver (and most people who carry such revolvers do not even carry spare ammunition), this feature seems unimportant.

On the negative side, the sights on the Webley M83 are hard to see because they're dark with a shallow, narrow back sight. Although the single-action pull is quite good, the double-action trigger is stiff and has a lot of slack. Additionally, the shooter must take care not to pinch his finger in the trigger, because it is quite easy to get a little flesh behind the trigger and pinch it as he pulls the last bit of the double-action trigger. The hammer spur would, I believe, be troublesome in pulling the weapon from a pocket; if the pistol belonged to me, I would cut off the spur, assuming it would still fire reliably. (With Colt revolvers, cutting the spur off sometimes creates a problem; you'd have to test to ensure that similar problems did not arise with the Webley.) Many Webley M83s have lanyard rings; my evaluation of this ring is that it merely adds bulk, and I would remove it if I carried an M83 in my pocket (however, it might be useful to secure the weapon to the wrist while sleeping in hazardous situations). Naturally, the recoil with this lighter weapon is heavier than it is with the same cartridge fired in a heavier weapon, and there is more muzzle flip, but these can be controlled enough for rapid double-action work on the cinema range. Because of its construction, reloading is slow, but keep in mind, also, how slow it is to reload a Colt SAA M1873.

The Webley RIC M83 .455 in full recoil.

These shortcomings have relatively easy solutions. The reloading problem can best be remedied by carrying a pair of the revolvers. Buy two of the 2 1/4-inch barrel models with nickel finish for maximum rust resistance. Never have both unloaded at the same time. Carry one in your pants pocket and one in your coat pocket. Use Webley Mk III manstopper loads (square wadcutter-style lead bullets with a cup point) and use standard military loads for practice. Paint your sights white to allow rapid pickup in poor light.

Right side of Webley RIC M83 .455.

Such a setup gives you a fine personal defense system that is lighter than Fitzgerald's cut-down .45 Colt New Service revolvers, weighs little more (if any) than a single Webley large-size .455 revolver, and in trained hands is more flexible than either. And remember that Henry Webley put 5 shots into an area 2 1/3 x 3 1/2 inches off-hand at 25 yards on May 19, 1884, with this revolver.

I am rather taken with the M83. Unfortunately, obtaining .455 can be difficult. Many M83 handguns you will encounter have had a hard life and

show the effects of 100 or more years of use. Also, black-powder cartridges and corrosive priming have adversely affected the bore of many. The lock work on the Webley M83 is sturdy, but parts, particularly the mainspring, will frequently crystallize with age.

Perhaps, like me, when you load up a Webley M83 and shove it in your pocket, you will suddenly hear the clopping of horse hooves on the cobblestones of a foggy London street or experience the sights and sounds (and smells) of a crowded Oriental bazaar . . . but even if you cannot be there, chances are your Webley has.

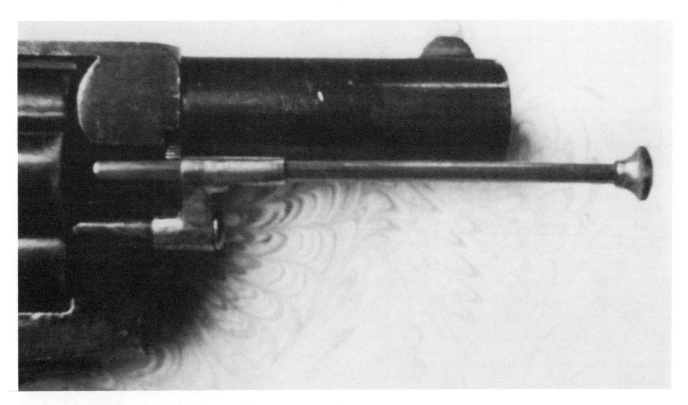

Reloading the M83 is accomplished by using the ejector rod.

The M83's cylinder is sharply fluted to reduce weight for pocket carry.

Famous Webley "winged bullet" can be seen on the frame of this M83.

Retailer mark on the Webley M83.

Webley Mk IV

11.43x19mmR
(.455 Webley Mk 2)

This is commonly known as the Boer War revolver. Adopted in 1899, the Mk IV continued as the standard British service handgun until 1913. NCOs and other ranks were issued this pistol, as was any officer who did not buy his own (British officers were allowed to buy their own handguns as long as they chambered a service cartridge). Most of these revolvers have a bird's-head grip and 4-inch barrel.

I found the grip to be less than satisfactory because it seemed to torque away to the left with each shot fired. The front sight is indistinct and dark, while the rear sight is dark and shallow. Applying white paint to either would allow them to pick up better in the darkened range and expedite indexing and triangulation.

The trigger pull on double action is heavy and long, while the single-action pull is light and crisp. The simultaneous case ejection is a wonderful feature. A mere push with the thumb, and the weapon pops open. Just pulling the barrel down causes the cases to fly out, allowing you to rapidly reload your weapon.

When you compare this weapon with the other military ordnance revolvers of its day, you quickly realize that the English soldier going off to the Boer War had a good piece of equipment. His counterparts in other European armies had small-bore revolvers or solid-frame rod ejection weapons. The grip is solid on the Webley Mk IV, even if it did torque away from my hands. Certainly, having a sideplate that swings forward (as on the Rast & Gasser) would make cleaning easier, but the Webley held up well in actual use. This well-made revolver has few parts of good materials, and the parts that do exist are big for the workload expected of them.

All Webleys remind me of locomotives, and like locomotives, they will get the long, difficult job done. The .455 Webley cartridge threw a

SPECIFICATIONS

Name: Webley Mk IV

Caliber: .455 Webley

Weight: 2.4 lbs.

Length: 11.25 in.

Feed: Cylinder w/ 6 chambers

Operation: SA or DA; top-break revolver

Sights: Front blade; rear notch

Muzzle velocity: 620 fps

Manufacturer: Webley & Scott

Status: Obsolete; common in Br. Comm.

.265-grain bullet at about 625 fps, although there were other lighter, faster loads. The British felt that a heavy, slow bullet gave better stopping power than a small fast bullet.

Many people who have never used the Webley Mk IV may view it as an obsolete weapon that cannot compete with a modern combat handgun.

Yet actual tests on the cinema range show that both it is an effective fighting tool. Great grandfather was not by any means ill-equipped to defend himself with a Webley Mk IV .455 revolver, and if you stumble upon one and are forced to use it, you will be in good company and better equipped than many military men in uniform today.

British troops carried the Webley Mk IV .455 during the Boer War.

Right side of Webley Mk IV .455 revolver, standard Boer War issue.

The Webley Mk IV's ejection system
is just one of the features that
make it a superior combat revolver.

Webley-Fosbery

11.43x19mmR
(.455 Webley Mk 2)

The Webley-Fosbery automatic revolver is an unusual weapon. There are only two or three automatic revolvers in the world, despite mystery writers' misrepresentations, and the Webley-Fosbery is possibly the best known. This weapon saw service during the Boer War and in the trenches of France during World War I.

The Webley-Fosbery presents a technological dead end, however, in comparison to contemporary weapons. It is well made and well designed, but it does nothing that justified the time and expense of its manufacture.

My test model was a standard .455 Webley-Fosbery. Throughout the test—both on the formal range as well as the cinema range—I encountered no malfunctions. This in itself is a credit to the weapons system and the designer, because the pistol is at least 70 years old. The sights were very hard to see because of the small,

---

SPECIFICATIONS

Name: Webley-Fosbery

Caliber: .455

Weight: 2.4 lbs.

Length: 11 1/4 inches with 6-inch barrel

Feed: Revolving cylinder

Operation: Recoil

Sights: Blade/notch

Muzzle velocity: NA

Manufacturer: Webley

Status: Obsolete

---

narrow rear notch and front blade. On the target range, it worked well; on the cinema range, it was hard to index it on target properly. Trigger pull is quite good, and, of course, this is a single-action-only weapon. Despite its being a revolver, it has to be cocked for the first round or carried cocked and locked. The pivoting safety is located on the left grip. I fired it at a formal target at 50 feet. Then just for curiosity, I also fired at a 14-inch metal plate at 70 yards. I compared the latter results with a Model 1911, Model 19, and SIG P210. Interestingly, with a Webley-Fosbery at 70 yards, I hit the plate only once out of six attempts. With all the other models, I was 100-percent accurate. So, obviously, the pistol must have been at fault, not the shooter. Of course, it could have been the loads I was using, newly manufactured Fiocchi .455 with 265-grain lead bullets, but I think it was a sight picture. The sights are very difficult to see on this pistol, and much too shallow, thin, and narrow for proper sight picture—at least for my eyes.

This was the first time that I had shot an automatic revolver, so I think that my comparisons above might be a little unfair. The proper way to compare the Webley-Fosbery is with a Mk VI Webley or a Colt Model 1911, both of which are big-bore World War I-vintage weapons. Both of these are far better practical fighting weapons.

On the cinema range, a couple of points came out that were rather interesting. First, you have to carry this weapon cocked and locked, which I suppose is acceptable. But that big Webley hammer cocked back there with its long firing pin on the hammer does give you a bad feeling, at least it did me. Without knowing more about how the locking system works on this pistol, I would be reluctant to carry it cocked and locked. That being the case, you end up with a weapon that you have to cock for the first shot because it does not cock itself when you pull the trigger mechanism, like a normal Webley revolver does. On the cinema range, I found that the best way to do this is to cock with the weak-hand thumb to avoid breaking your grip when you have it in a two-hand hold. Second, recoil is roughly the same as on a Mk VI Webley in the same caliber. I found this interesting because I thought that it would be a little less.

The grips are good, although the weapon did tend to shoot low on the cinema range. The cylinder can be opened while the weapon is cocked and the safety engaged, which allows you to reload the weapon rapidly. And you still have a cocked weapon available, an advantage if you're attacked and you must advance forward with your pistol cocked and the safety on. You can empty your cylinder and reload rapidly without dropping the hammer.

These pistols were typically loaded or used with the Prideaux speed loader, which was quite common during World War I. With practice, a soldier could load his weapon as rapidly as any magazine self-loader. The Prideaux speed loaders each have a small ring on them for the purpose of attaching to a string, which the soldier then tied to his uniform so that he could add three to five Prideaux speed loaders. As he needed them, he added the speed loaders and as soon as the cylinder was filled, he dropped them. When the battle was over, the soldier could retrieve his speed loaders.

The weapon, though it felt good in my hand, also felt strange and slightly unsafe—due to my unfamiliarity with it and concern about the metal parts fatiguing and causing the safety to fail. The safety fell readily to hand, and it could be disengaged with your thumb without breaking your grip, but, unfortunately, it could not be reapplied without breaking your grip.

All in all, it is hard to see how this pistol is really better than the double-action Webley Mk VI. It does not fire any faster than a double-action revolver; it has to be cocked for the first round, unlike a double-action revolver; it loads no faster than a double-action revolver; and it is mechanically more complex than a double-action revolver. It is probably slightly better than many automatic pistols of the early 1900s, but it was not as good or as useful after the Model 1911 Colt came out.

My shot group at 50 feet was 2 3/4 inches for four of the five rounds, with one flyer. The 2 3/4-inch groups show that the weapon is capable of fairly good accuracy. I attribute the flyer more to the sight problem than intrinsic weapon inaccuracy. Walter Winans, a noted English target shooter of the Edwardian period (early 1900s), used a Webley-Fosbery and was able to shoot it quite rapidly on the normal revolver courses of his day.

Even though it is a technological dead end, it is still a quite interesting and clever design.

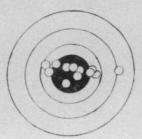

An illustration of the test target shot by noted English shooter Walter Winans with the Webley-Fosbery around the turn of the century.

The 50-foot test target shot by the author using a Webley-Fosbery.

Right side of Webley-Fosbery automatic revolver .455.

The Webley-Fosbery can be carried with the hammer down and the safety applied, which prevents cocking and firing the weapon as illustrated here.

Cocked-and-locked carry of the Webley-Fosbery. Note that the safety is applied.

Close-up view of Webley-Fosbery safety and firing mechanism.

Cylinder showing zigzag pattern that allows functioning of the Webley-Fosbery.

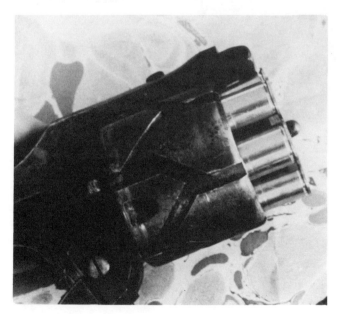

The Webley-Fosbery has the standard Webley extraction system.

The British World War I-era speed loader that could be used with Webley-Fosbery or Webley revolvers. Such a device permitted a rate of fire rivaling that of an autoloader in the trenches.

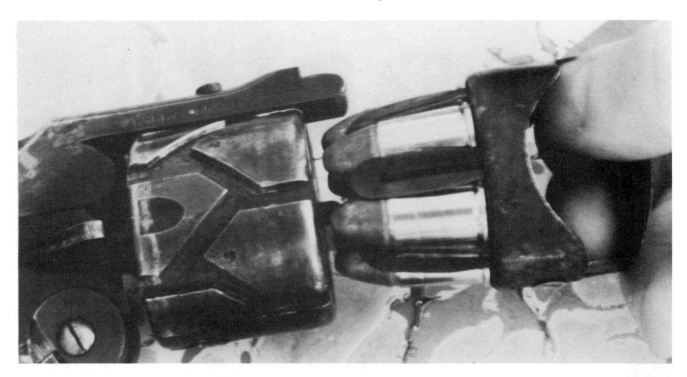

Inserting a speed loader into a Webley-Fosbery cylinder.

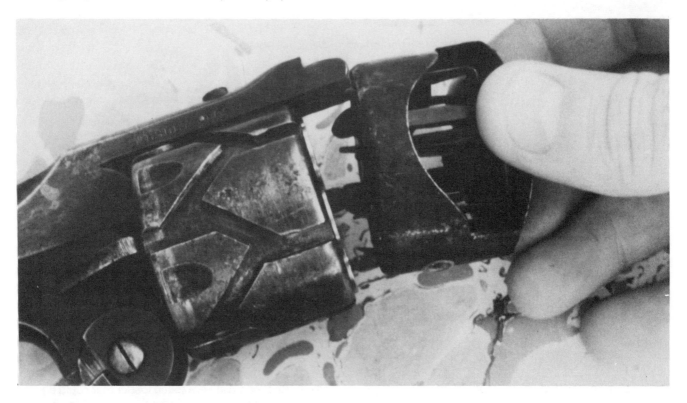

After the cartridges were firmly seated in the weapon, the speed loader could be dropped to the ground for later retrieval or, as was more common, the loaders were attached to the shooter's vest by cords.

Webley-Wilkinson 1911

11.43x19mmR
(.455 Webley Mk 2)

In the pre-World War I British army, the government supplied revolvers to noncommissioned officers, but commissioned officers were supposed to supply their own. As mentioned earlier, officers' weapons were supposed to chamber the government cartridge (.455), but this requirement was frequently winked at, which is why .577 pistols are encountered from time to time.

Officers tended to come from the upper classes or at least upper-middle classes, and they frequently had nice pistols. I imagine everyone who reads this material is familiar with the Lanchester four-barrel pistols chambered for .455. These were obviously designed for the officer who did not trust mechanical devices like revolvers, much less autoloading pistols.

The revolver I tested was manufactured to the specifications laid down by the famous Wilkinson Sword Company—no doubt, many a young officer

SPECIFICATIONS

Name: Webley-Wilkinson 1911

Caliber: .455 Webley

Weight: 2.4 lbs.

Length: 11 1/4 in.

Feed: Revolver

Operation: DA

Sights: Blade/notch

Muzzle velocity: NA

Manufacturer: Webley

Status: Obsolete

picked a sword and a pistol on one visit. The Webley-Wilkinson 1911 has all the good features of the Webley Mark VI, only better fitted and finished. It had a beautiful blue finish, much nicer than anything I have previously had the opportunity to use, and was finely fitted. The trigger was smooth-faced and well fitted. The trigger pull on double action was heavy but smooth and had no stacking. The single-action trigger broke like a glass rod. Wonderful!

The rear sight had a wide, shallow V that allowed for quick indexing. The front sight had a small bead, making quick indexing slower but acceptable. It would be better if it were painted white.

Reloading with this pistol is quick, as it is with the standard Webley. The grip feels good in the hand, does not shift with rapid repeat shots, and is excellent for work on the cinema range. The example tested had an improperly fitted (or worn) locking bolt, and the cylinder thus suffered from misalignment. Groups were larger than I would have expected. This may have resulted from ammunition not perfectly fitted to the barrel (I was using commercial Fiocchi loads) or a pitted bore combined with the bad lock-up, but since it was not my pistol, I could not get it repaired or take the time to measure the bore. It was disappointing, however.

Fairly light in weight and reasonable in size, the Webley .455 cartridge is not a magnum by any means, but I would not hesitate to use this weapon for self-defense if it shot better groups. The British officer who purchased this weapon made a good selection, and the men who designed and made this pistol were real craftsmen. They may be long gone to their reward, but the pistol they made still speaks eloquently of their skills.

The author test-firing the Webley-Wilkinson, a beautiful weapon even now!

The top strap of the Webley-Wilkinson combat revolver.

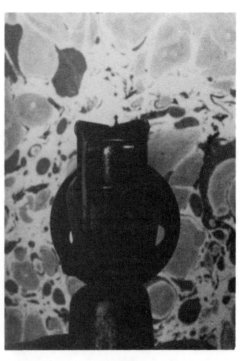

No grip adapter is necessary with the Webley revolver, which has great factory grips.

The sights on the Webley-Wilkinson allow rapid pickup in dim light.

The 50-foot test target fired with the Webley-Wilkinson .455.

Webley Mk VI

(.455 Webley Mk 2)

If the Webley Mk IV was the Boer War handgun, the Webley Mk VI was the standard British handgun of World Wars I and II. Obviously, there was an acute shortage of revolvers in World War I, and many different handguns (including the Webley Mk IV) from many countries were used, but effective May 1915, the Webley Mk VI was the standard.

It differs from the earlier Mk IV in that it has a front sight that can be replaced, its 6-inch barrel is 2 inches longer, and it uses the grip style from the commercial Webley revolver series.

The rear sight on the Mk VI was about the same as that found on the Mk IV and is too shallow and dark. The front sight is better since it is more distinct on the cinema range and also easier to hold elevation and windage on the formal range. The longer barrel, to my mind, is actually a drawback, and the Royal Irish Constabulary thought so as well: when it bought some com-

SPECIFICATIONS

Name: Webley Mk VI

Caliber: .455 Webley Mk 2

Weight: 2.4 lbs.

Length: 11 1/4 in.

Feed: Revolver

Operation: DA

Sights: Blade/notch

Muzzle velocity: NA

Manufacturer: Webley

Status: Obsolete

mercial Mk VI pistols in early 1920, it specified a 4-inch barrel. However, it does not seem to balance poorly and, except as a concealment piece, the 6-inch barrel is not a major weakness.

The grip is what really makes the Webley Mk VI the best .455 Webley ever made. It is wonderful. Unlike the Colt and Smith & Wesson revolvers, you can use a Webley Mk VI without any adapter. It will not shift in your hand while firing rapid double-action strings.

The ejection system on the Webley Mk VI is as good as that found on the earlier Webley pistols. By using the Prideaux speed loaders in conjunction with the rapid ejection system, you can almost equal the rate of fire of the M1911.

During World War I, the British also adopted a bayonet and stock for the Webley Mk VI. At first, both of these features stuck me as rather silly because they impair handiness, which is one of the prime features of a handgun. In actual use, the two items are quite serviceable. I found with the Webley, as with all stocked handguns, that the accuracy was not really much better than with a good, two-hand Weaver stance; but for a less-experienced shooter or when one is tired and out of breath, the stock can be truly useful. At first, the bayonet seems worthless, but when you put it on the pistol in the stock, you get a dandy little carbine. Keeping in mind that there was no submachine gun until 1918 and none in the British services, a stocked Webley Mk VI with bayonet seems just the ticket for trench clearing. It is light, easy to shoot, and quick to reload. Power for close-range trench work is satisfactory, and a bayonet is a dandy little tool for keeping prisoners in line. (When I was in the U. S. Army, I stabbed a guy with the bayonet fixed to my M14 rifle, and I remember how easily it went in when I had the weight of the weapon behind it and how well behaved he was thereafter.)

I got the stock and bayonet simply because they are featured accessories in the various Webley books, and I thought they should be tested. As with other weapons, when I got the Webley Mk VI to test, I really did not expect much. But once you get it in your hands, you quickly realize that you have a real fighting tool.

Because the sun never set on the British Empire until after World War II, you are likely to find a Webley Mk VI just about anywhere. I suppose they might be rare in Latin America (except for the Latin copies), but they are common in the Middle East and Africa as well as all over Europe.

Author test-firing the Webley Mk VI .455 revolver with and without stock (above right and right).

Many that we see in the United States are rechambered for the .45 ACP round, which uses a .451 bullet. Projectile fit thus is loose, and accuracy suffers. I realize that you may not need many rounds, but those you need should be accurate. This points up the worst feature of the Webley Mk VI, namely that .455 Webley ammuni-tion is frequently in short supply. Still if you can get the ammunition, you will have a solid companion, one that never failed anyone in the muddy trenches of France in 1916 and that is unlikely to abandon you in less vigorous circumstances.

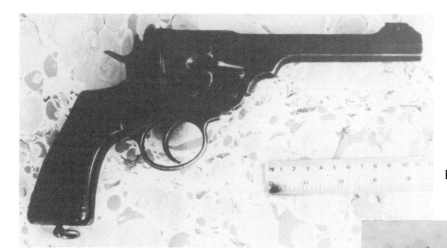

Right side of Webley Mk VI .455 revolver.

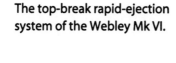

The top-break rapid-ejection system of the Webley Mk VI.

Webley Mk VI .455 revolver with the 50-foot test target.

Webley
Self-Loading Mk 1

11.43x23.5mmSR
(.455 Webley Auto Pistol)

The Webley self-loading Mark 1 pistol chambered for the .455 SL cartridge was an early rival of the Colt Government Model. During World War I, it was commonly found in the hands of officers who purchased it privately. One hundred were also issued to the Royal Horse Artillery (50 with a shoulder stock) and to the Royal Flying Corps for one year, where it was used for shooting at enemy airplanes in the initial stages of the war. As the ".455 Pistol, Self-Loading Mark 1, 1912," it was adopted as supplemental issue in the navy and ended up most commonly found in the hands of naval landing parties. Although less rugged than the Colt and subject to mud clogging it (no doubt, due to the action opening up at the top, exposing the entire interior to the elements), it was easily stripped and well made of good materials.

The example tested was manufactured in 1913

SPECIFICATIONS

Name: Webley Self-Loading MK 1

Caliber: 11.43x23mmSR

Weight: 2.4 lbs.

Length: 8 1/2 in.

Feed: In-line box

Operation: Recoil

Sights: Blade/notch

Muzzle velocity: NA

Manufacturer: Webley

Status: Obsolete

and bore serial number 870. It was marked with British government acceptance stamps. Despite its age, its condition is still quite good, and I encountered no malfunctions. On the formal range, I found it difficult to hold the elevation, due to the wing-like projections on the rear sight. These made it difficult to keep it level and resulted in larger than average groups of 7 1/8 inches.

Recoil is heavier than that found with the Colt Government Model, which I found surprising. One interesting feature is that with the slide locked open, as occurs after the last shot, inserting a fresh, loaded magazine will cause the slide to close instantly, thereby loading the weapon. All semiautomatic pistols should adopt this feature because it speeds up the loading cycle and has no drawbacks that I can see. One amusing feature is the two-step magazine that allows you to withdraw the magazine and use the pistol for a single shot firing, reserving the magazine for an emergency. Such cutoffs were common at the time with rifles, but I would be surprised if anyone ever really used this feature in the field.

On the cinema range, I quickly zeroed in on the slender diameter and light barrel (due to the large hole down it) as the cause of the balance point being different from that on the Colt Government Model. The Webley balances off the grip instead of in front of it, as on the Colt. I found that I was able to hit quickly and accurately because the high front sight and wide rear sight allowed rapid indexing. The same wide rear sight with wings that made it difficult on the formal range made it fast on the darkened cinema range. For picking combat handguns, I would prefer a good performance on the cinema range to that on the formal range. I noted also that the light-colored firing pin of the Webley, at the rear of the slide, also aided front-sight aligning for rapid indexing. Of course, painting the sights white would even be better.

Despite criticisms in a variety of books about the straight grip of the Webley, I found it of no concern. It might look awkward, but it did not shoot awkwardly. Perhaps the other commentators never shot a Webley self-loader on a true

combat course, if they shot one at all.

On the other hand, I found the lack of a safety, other than a grip safety, quite disturbing. You are reduced to carrying it empty and pulling the

Author firing the Webley .455 autoloader.

British cavalryman in World War I with Webley self-loading pistol.

I suppose for the British Marine or officer with his private-purpose Webley in World War I, this would not be as big a problem as it is for me, but I think it too dangerous cocked for the rough and tumble of the trenches and too slow if not cocked. Overall, it has good stopping power and reloading ability, but I prefer a Webley Mk VI with speed loaders, which is equally fast to shoot, equal in power, and much safer to handle.

The magazine on the Webley self-loading Mk 1 has two positions. By lowering it to the second notch, the shooter can use the weapon as a single loader and reserve the magazine for an emergency. This reflects fuzzy thinking on the part of the gun's designer.

slide to the rear, loading it when needed. I find this unacceptable because I may need my other hand at a critical time to do something else. At the least, it is slow and causes you to break your grip, slowing up response time considerably. The other option is to carry it loaded with the hammer down, but you run the risk of an accidental discharge if you drop the weapon. Last, you can carry it cocked and depend on the grip safety. If the grip safety spring were stronger, that would be fine, but the Webley spring is so light that the least touch disengages it. I did find it was easier to avoid depressing it when the pistol was shot with one hand rather than with my typical two-hand hold; the weak hand tends to push the pistol into the strong hand, instantly disengaging the safety on the Webley.

The 50-foot test target shot with the Webley .455 autoloader.

Left side of Webley .455 autoloader.

Smith & Wesson New Century (Triplelock)

11.43x19mmR
(.455 Webley Mk 2)

The famous Smith & Wesson Triplelock revolver was designed to be a target shooter and hunter's weapon, not a military weapon. Between September 1914 and September 1916, the British bought 5,000 S&W Ejector First Model, and 69,755 Second Model revolvers chambered for the Mk 2 .455 cartridges. The mud of France was the undoing of the design and no doubt a fitter's nightmare. However, anyone who really likes revolvers and sees a Triplelock wants one. I know I would truly love to get a target-sighted, 5-inch-barrel .44 Special Triplelock. Every time I see a picture of Elmer Keith's old Triplelock, I get green with envy.

My favorite story involving a Triplelock .455 was the one related by Elmer Keith. In it, a bear got into a packtrain, causing much confusion. A man in the train yanked off his glove with his teeth, pulled his .455 Triplelock with heavy handloads, and placed two rounds into the bear as it ran toward them.

SPECIFICATIONS

Name: S&W New Century Triplelock

Caliber: 455

Weight: 2.4 lbs.

Length: 11 3/4 in.

Feed: Revolver

Operation: DA

Sights: Blade/notch

Muzzle velocity: NA

Manufacturer: Smith & Wesson

Status: Obsolete

Great story!

I looked forward to testing this revolver and thus was disappointed at the poor group I fired. A 3 1/8-inch group really is too big, but I think it was because of the sights, not the weapon. The sights were hard to see on the cinema range because they were dark, and the rear notch was low and small. Obviously, the sights would be better if white and broader. The trigger in the double-action mode was quite good and allowed rapid follow-up shots. With the low recoil of the .455 cartridge, the non-Magna grip design was not much of a problem, unlike the problem it has when a high-powered .44 Special load is fired. Only with rapid repeat shots did the .455 rounds drive the weapon into the web of the hand enough to cause shifting. Interestingly enough, the face of the trigger was smooth, like that of the modern combat revolver, unlike the grooved triggers that you would expect with a revolver for target shooting. Smooth triggers are better for fast double-action shooting.

This is a large pistol with an N-frame and 6 1/2-inch barrel, but the skinny barrel and large holes in the barrel and cylinder prevent the weight from being excessive. Although the New Century is a very well-made and designed revolver, better revolvers exist for the 1914 period. Its grip design is not as good as that found on a Webley Mk VI, which in my estimation is a better military revolver for an officer or NCO. If the sights were better, the barrel shorter (so the overall length would be less), and the grips better, my evaluation might be different.

Many of these .455 Webley-caliber, British Smith & Wessons were reimported to the United States as surplus and were rechambered for .45 Long Colt or had the rear of the cylinder milled off to permit the use of .45 ACP rounds with half-moon clips. Both these conversions were workable but not entirely satisfactory: the cham-

The author firing Smith & Wesson .455 Triplelock.

British imperial forces practice with their handguns in the desert. Both Smith & Wesson and Webley .38 and .455 are present in this photo.

ber profile of the .455 round is larger than that of the .45 Long Colt, leaving a slightly oversize chamber, and when the rear face of the cylinder was milled to accept half-moon clips, the extractor star was also thinned appreciably, thus weakening it.

Right side of Smith & Wesson Triplelock .455.

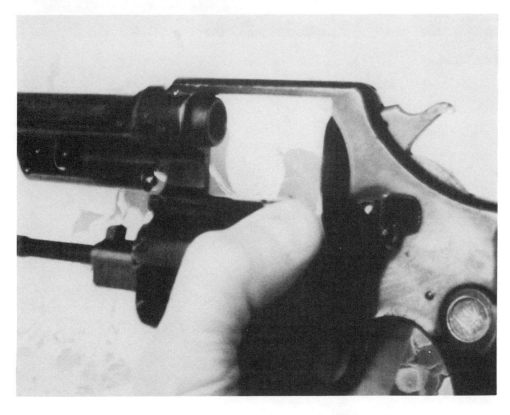

The extra locking feature that gave the Triplelock its name.

The 50-foot test target shot with the Triplelock.

Old Pattern 5-Inch No. 2 Mk 1, M1915

Various Calibers

My first big-bore pistol was one of those Spanish-built Webley copies. At the time, being only 10 years old, I did not realize how rare they were. During World War I, Britain was in desperate need of small arms. She bought them at home, remodeled obsolete weapons, allowed her officers to buy whatever they could find, shopped in the United States, and even went to Spain to find an additional manufacturer.

The Old Pattern No. 1 and No. 2 Mk 1 revolvers were made by Garate (No. 1) and Trocaola, Aranzabal y Cia (No. 2) and adopted by the British army in November 1915 as a substitute standard. Revolvers by these two makers, and which bear British broad-arrow ownership stamps, are serviceable if in good condition. They were a hybrid design that incorporated a Webley-style top-break action and a Smith & Wesson-type action lock, and they had a semi-

SPECIFICATIONS

Name: Old Pattern 5-Inch No. 2 Mk 1

Caliber: 455

Weight: 2.4 lbs.

Length: Approximately 10 in.

Feed: Revolver

Operation: DA

Sights: Blade/notch

Muzzle velocity: NA

Manufacturer: Garate & Trocaola Aranzabal y Cia

Status: Obsolete

bird's-head grip common to neither. This weapon is also known as the "Hermanos Ovbea" and was made in 10mm Italian for the Italian market and 44-40 W.C.F. for the South American market.

My pistol is a 44-40 W.C.F. I fired it until the top latch broke, and then I drilled it and put a bolt and wing nut in the hole to hold it down. Finally, when the screw holding the barrel onto the frame broke, I put a hardware store bolt through it. About the time I made this last gunsmithing addition, it dawned on me that perhaps this was not such a fine weapon, and I retired it in favor of a Smith & Wesson .38 revolver. I was 12 at the time.

Although these pistols were obviously made to rather lax standards of finish, the one I tested functioned without a hitch. My accuracy was not great at 6 1/16 inches, but the sights were very hard to see, and my left-hand middle finger was wrapped on the knuckle, which did not help my performance. The worst feature of this weapon is the lack of a locking bolt. When the cylinder is empty, it will properly rotate; when it is full and you fire three rounds, it fires fine. Then when you pull the fourth shot, the weight of the loaded cylinder chambers is such that the cylinder falls to the bottom of the frame, forcing the empty chamber to come up again. A bolt lock would prevent this. This problem could easily get you killed when you jump into a trench, pull out your pistol, and start firing. You end up with a 3-shot revolver because of this design flaw. (This feature was shared by the Type 26 Japanese revolver and was quite annoying with it as well.)

On the cinema range, the front sight was quite thin and hard to see. It needs to be wider and painted a contrasting color or at least white. The rear sight was totally impossible to see and needs to be wider and with contrasting color. The design shared the good feature of all Webleys and it allowed very fast reloading. Merely flipping the latch and breaking the frame down would flip out all the empties. The smooth finish of the trigger allowed rapid repeat shots, and the

Right side of Old Pattern 5-Inch No. 2 Mk 1 M1915 .455.

Left side of Old Pattern 5-Inch No. 2 Mk 1 M1915 .455.

grip felt good when used on the instinctive cinema range program. The trigger pull was heavy but acceptable. The low recoil of the .455 ammo made repeat shots easy.

The British purchased many thousands of these pistols as a substitute standard in World War I, yet you rarely see them. While not as well finished nor designed as a Webley, it is still better than .32 ACP caliber pocket pistols manufactured in Spain and sold to France during World War I. My real questions is, where did they all go? Maybe you will find them in your travels.

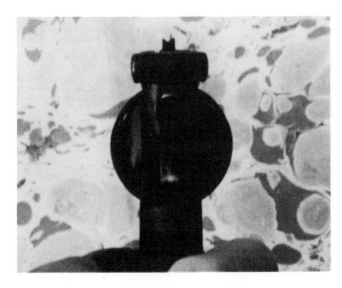

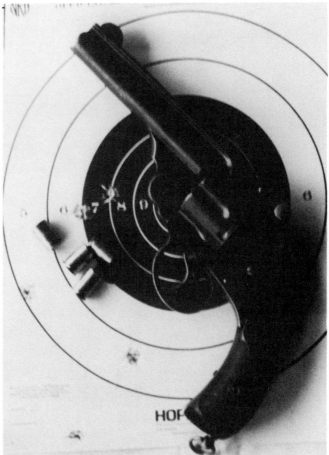

The rear sight on the Old Pattern is not nearly as good as those on the Webley series.

The 50-foot test target shot with the Old Pattern 5-Inch.

Webley
Mk IV

.38/200

Although the Webley Mk VI .455 had come out of World War I with a sterling record for reliability and power, the postwar British army leaders decided that it was too heavy and had too much recoil. Somehow they managed to convince themselves that a .38 Smith & Wesson cartridge loaded with a 200-grain bullet at 750 fps would perform just as well, while at the same time reducing the recoil and weight of the weapons.

The Webley firm offered its Webley Mk IV model. It had all the good features of the Webley Mk VI and, in simple terms, might be viewed as a scaled-down Mk VI. However, after years of testing and modifications, the British military decided to adopt Enfield-designed revolvers instead.

Having incurred heavy development expenses, the Webley firm sought to recoup its losses by offering the model to British police forces. Many purchased them, and they were still commonly

SPECIFICATIONS	
Name:	Webley Mk IV
Caliber:	.38/200
Weight:	1.7 lbs.
Length:	10.2 in.
Feed:	NA
Operation:	Manual, revolver
Sights:	NA
Muzzle velocity:	595 fps
Manufacturer:	Royal Small Arms, Enfield
Status:	Obsolete

found among police agencies when I first visited London in 1978. Similarly, overseas police agencies under British influence frequently purchased the Webley Mk IV .38.

Whether the Webley firm would have ever recouped its money from police sales remains questionable. With the advent of World War II, however, the Enfield factory was unable to produce its handguns fast enough, and a call went out to Webley. The firm answered the call, as might be anticipated, by providing high-quality weapons in the latest available pattern for the then-service cartridge of .38 Smith & Wesson.

From the beginning, it was obvious that the Webley was better finished than the Enfield revolver, and the difference was even more marked as the war progressed. It is only fair to state, however, that many late-war weapons from both factories are better than some current production revolvers. Webley, ever sensitive to its public image, marked its pistols "war finish" so that the soldier in the field would understand why the high-polish blue was missing.

The Webley Mk IV .38/200 can be viewed as a Webley Mk VI that was left out in the rain, basically. The comments I made pertaining to the Webley Mk IV apply here as well. The grip is very good, and no adapter is needed. The trigger pull on the double action is somewhat heavy in comparison to a Smith & Wesson of the same period but generally quite smooth. Single-action pull is very satisfactory. The sights are quick to use on the darkened range and allow quick indexing. The top-break feature allows you to reload rapidly and with brute motor skills rather than the finer motor skills required with Colt or Smith & Wesson revolvers.

I am quite fond of these revolvers except for one bad feature: the caliber. The .38 Smith & Wesson as commonly found is really on the same level as a .380 ACP—hardly inspired. Given a choice between a Webley Mk VI .455 and a Webley .38/200, I would opt for the former. Neither load, of course, will penetrate body armor or full magazine pouches, so for today's active military person they are lacking. For individual defense work this is not as much of a problem.

As a result of the dispersion of British troops during World War II, these revolvers are likely to be found in most parts of the world, except South America. Frequently, their flaking finish and top-break action render them undesirable and quite low in price. There is some collector interest is them, but enough mint-condition specimens abound that you can still find shooting examples at very low prices. Should you need a weapon and encounter a Webley Mk IV .38 in some backwater area such as the Central African Republic or California, snap up this well-tested and developed combat weapon. Select, if possible, the 200-grain bullets, which tend to tumble and thereby give better stopping power. But even with standard ball 172 FMJ loads or commercial 146-grain loads, I would rather have a Webley (or Enfield) .38 than almost any .380 pistol made (not taking concealment into consideration).

Webley Mk IV .38/200 being tested.

The rear sight on the Webley Mk IV .38/200.

Webley Mk IV .38/200 with 50-foot test target.

Enfield Revolver No. 2 Mk 1

9x20mmR
(.38 Smith & Wesson)

I tested two Enfield Revolver models: No. 2 Mk 1, the original model with a hammer spur with both a single-action and double-action trigger mechanism, and No. 2 Mk 2, known commonly as the Commando Model but actually introduced for tankers, who found that the hammer spur on the original Enfield Revolver would catch on things inside the tank. Thus, the hammer spur was deleted, the single-action notch left off the hammer, and the mainspring lightened.

The Enfield Revolver was developed in the 1920s as a replacement for the Webley Mk VI revolver. The Webley Mk VI had served honorably during World War I and proven itself to be reliable and quite a good stopper. But when the war was over, the British military wanted a lighter, smaller, easier-to-handle weapon with less recoil. After much experimentation, the British convinced themselves that the 200-grain, .38 Smith & Wesson revolver round

SPECIFICATIONS

Name: Enfield Revolver No. 2 Mk 1

Caliber: .38 (.380 in revolver)

Weight: 1.6 lbs.

Length: 10.25 in.

Feed: Cylinder w/ 6 chambers

Operation: SA or DA

Sights: Front blade; rear square notch

Muzzle velocity: 600 fps

Manufacturer: Enfield

Status: Obsolete

could achieve this result. Interestingly enough, after having adopted the 200-grain load, the British decided that these were in violation of the Hague Peace Conference, so they went to a full-metal-jacketed 178-grain bullet. As a result, most of the weapons that you will encounter are sighted for those loads.

When you use commercial ammunition that has 148-grain lead bullets, it shoots to a different point of impact. This is because the bullets are considerably lighter than the British wartime loads. Front sights are easily replaced, and you can cause the point of impact to be raised by changing the front sight. The problem with this pistol, of course, is the load: .38 caliber, 178-grain, full-metal-jacketed loads at 650-odd fps simply are not reliable manstoppers. Throughout World War II, this was the major complaint about the Enfield. That is only reasonable when you consider that these loads are slightly less powerful than .380 automatic loads.

During World War II, the revolvers were finish designed and manufactured in a variety of places: the Royal Ordnance factory at Enfield and Albion Motors near Glasgow; and the Singer Company near Glasgow made component parts, which were shipped to Enfield. They are also made in Australia.

Quality varies tremendously with some of the weapons. There is also a variety of painted finishes existing, again of varying quality. One model I tested had been manufactured in 1933 at Enfield; it was a well-finished, well-made weapon with wooden grips. But I have seen wartime weapons with components made at the Singer factory, and they were very poor production weapons—certainly nothing to brag about. Manufacturing revolvers, quite frankly, does not lend itself to inexperienced workers and rushed situations.

On both of these models, the grips feel very good in my hand, and unlike with the Smith & Wesson revolvers, no adapters are really necessary to get a good grip. Recoil is quite low, so I suppose the British achieved their goal of producing a low-recoil weapon. It feels much like a .22 in my hand. The bullet holes are very hard to see because the bullets themselves cut very small holes in the target material. I suppose the result would be the same whether shooting people or other targets. The double-action pull is heavy, tends to stick quite a bit, and is certainly harder than a Smith & Wesson double-action pull. The single-action pull on the hammer model is adequate, but not spectacular by any means.

The report of the .38/200 round is also quite low and not very much greater than that of a .22. The notch of the rear sight is too narrow, although the sight picture is quite good. The front sight is quite narrow also and tends to reduce one's ability to shoot accurately. The cylinder latch breaks with some difficulty, and there is a tendency to hit the latch with your thumb and tie up the weapon. However, once it is opened, it does eject the empty cases vigorously, allowing you to reload rapidly. I am not certain if a speed loader is available for these revolvers, but I have never seen one. Certainly

without a speed loader, you are losing the benefit of the rapid-ejection systems.

The groups with the No. 2 Mk 1 Enfield ran about 2 3/8 inches at 50 feet. With the No. 2 Mk 1 model, I got slightly larger groups of 3 inches. All in all, that is a pretty good performance for these weapons; with the Model 19 Smith & Wesson I got about 2 1/2-inch groups that day.

On the No. 2 Mk 1, the double-action pull is also quite heavy, and it stacks quite a bit at the end. The sights, again, are hard to see, and the grips on this pistol are plastic. Unlike the wood-grip 1933 double-action model, the hammerless model has thumb indentations that were supposedly designed to align the weapon for rapid double-action work. Many stories have been told about the Commandos running through courses, pointing these weapons at man targets and firing rapid bursts. The weapon can be fired rapidly in bursts of two because of the light recoil and low report. The heavy double-action pull makes it difficult to stay on the target while you are pulling the double-action trigger, because of the weapon's lightness and the stacking of the trigger.

The grip, however, feels quite bulky in my hand. It tapers in the wrong way—it gets wider at the bottom—which causes you to have a poor overall grip. Grips for double-action revolvers should be designed like your hand when you close it. They should get smaller at the bottom. Grips that are bigger at the bottom tend to spread your hand out and weaken your grasp rather than strengthen it. If you really want to have the proper grasp for double-action work, it should be strong, just like closing your fist.

These Enfields have been left behind all over the world by the many Commonwealth armies and police forces, and it is likely you will encounter them in your travels. The pistol is accurate enough and its ejection system is positive, but the cartridges themselves are low in power. The pistols are not easy to conceal and are lightweight (only slightly lighter than a modern Smith & Wesson K-frame revolver). In a better caliber, this weapon might be acceptable, but in this caliber, it seems doubtful to me. There are better revolver choices available.

Right side of Enfield
Revolver No. 2 Mk 1.

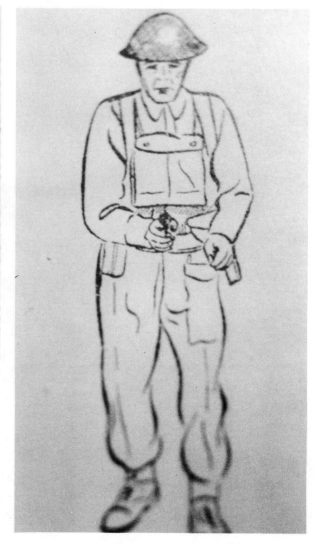

The grips of the early Enfield .38 revolver (above left)
contrasted with later World War II grips of a similar design
(above right). As the war progressed, the British turned the
Enfield into more of a weapon and less of a target tool, as is
evidenced by the illustration from a British wartime military
manual (right) demonstrating how crack commando units
should use the handgun aggressively.

British armored troops adopted the Enfield revolver with the spurless hammer.

The Enfield Revolver No. 2 Mk 1 with 50-foot test target.

Smith & Wesson Military and Police

9x29mmR/9x20mmR
(.38 Special/.38 Smith & Wesson)

This pistol design is the most successful revolver ever made. Although more commonly found in civilian hands, especially for law enforcement tasks and personal defense, it has also been adopted by many military organizations. During World War II, the British and Commonwealth countries used almost 900,000 of these revolvers in .38 Smith & Wesson caliber. The consensus during the war was that the Enfield was more durable under wartime conditions, but that the Smith & Wesson was better finished and more accurate to shoot. The British military and police (M&P) revolvers had 4-, 5-, and 6-inch barrels and came in blue and sandblasted finish. They are likely to be encountered throughout the British or Commonwealth wartime area of operations. Some 1,125,000 No. 2 (M&P) and "victory" model S&W receivers were issued in World War II to British and Commonwealth fighting men.

SPECIFICATIONS

Name: S&W Military & Police

Caliber: British Service .380 (.38 S&W)

Weight: 1.8 lbs.

Length: 10.2 in.

Feed: Cylinder w/ 6 chambers

Operation: NA

Sights: Front blade; rear square notch

Muzzle velocity: 600 fps

Manufacturer: Smith & Wesson

Status: Obsolete

As mentioned previously, the British had substituted the .38 Smith & Wesson cartridge with a 200-grain bullet for the .455 but later decided that violated the Hague Conference agreements. Instead, they adopted a 172-grain, full-metal-jacketed bullet and thus got a .380 ACP equivalent, which was not sufficient. To make matters worse, although the British M&Ps are sighted for the 200-grain bullet, the most common commercial bullet load available is 145-grain. Such loads do not shoot to point of aim with handguns sighted for 200-grain bullets.

Exactly how the British managed to convince themselves that a 200-grain .38 was equal to a 265-grain .455 has always been unclear to me. Apparently, the 200-grain bullet would tumble when it hit flesh, thereby improving lethality and stopping power. Interestingly enough, the first known shooting incident involving these bullets took place 8 miles from where I live. In

1927 in East St. Louis, Illinois, Patrolman Edward Sweeney shot a robber with this load. I imagine that Mr. Sweeney was much more satisfied with the results than many British soldiers in later years.

U.S. forces also used the M&P during World War II, except their guns were chambered in .38 Special and used the 130-grain full-metal-jacket loads to comply with the Hague Conference. U.S. guns are either 2- or 4-inch guns and are typically found with a parkerized finished and a lanyard ring in the butt. Many were used by pilots during the war, especially in navy and marine units, and they were also common with defense installation guards and military intelligence units.

The M&P revolver is an excellent piece of equipment, but it has a few shortcomings. Without an adapter, the grips do not feel good in my hand (but that is easily remedied), and the weapon is not as easy to clean as a Lebel or Rast & Gasser. The rear sight is low and narrow, which made triangulation difficult.

On the plus side, the parts are good-size and seem to last a long time unless abusive ammunition is used. If anything, they get better as they are used. Because of the cartridge used and the all-steel construction, recoil is low, and thus rapid double-action fire is easily achieved. Single-action pulls are good, and the double-action pull, although longer than on a newly made model (pre-1945 guns all have the "long action"), are

quite good with little stacking. Sights are distinctive in the front, although painting them white would improve them. Even so, they can be rapidly indexed. The 4-inch weapons are quite handy and can be carried easily concealed. The 2-inch model is easier to shoot than a J-frame 2-inch because of the width of the sights and the trigger pull, which is always better on a K-frame with its leaf spring.

Before this test, I had always considered the M&P to be heavy for its bore size and barrel length, especially since you could get the same power and barrel length in a J-frame-size pistol. However, in my tests on the cinema range, the M&P 2-inch was so much easier to shoot than the J-frame—because of both lower recoil and better indexing—that I became a convert. As long as you do not try to make a 2-inch M&P into a holster gun (rather, carry it in a pocket in an overcoat or of fatigue shirt), it makes a handy military handgun.

A 50-year old M&P revolver stands as ready to answer the call to duty today as it was the day it left Springfield, Massachusetts. Except for minor cosmetic changes, the weapons today are no different than the weapons made when Hitler was still on the scene. The British Tommie in Burma in 1945 with his Smith & Wesson M&P was well prepared, and if you are offered one in Rangoon next month, you will be also. These handguns are widely available.

Smith & Wesson Military and Police .38/200 with 5-inch barrel.

The 50-foot test target shot with the Smith & Wesson Military and Police 5-inch barrel.

Luger M23
7.65x22 (.30 Luger)

SPECIFICATIONS

Name: Luger M23

Caliber: 7.65mm

Weight: 1.9 lbs.

Length: 8 3/4 in.

Feed: In-column single row

Operation: Recoil, toggle lock

Sights: Front blade; rear notch

Muzzle velocity: 1,250 fps

Manufacturer: DWM

Status: Obsolete

Finland was part of Imperial Russia until the end of World War I, when it managed to get its independence. Still, the Finns faced a powerful enemy, the new Soviet Union, which viewed Finnish territory as its own and would have liked access to the Baltic that Finland would give them. The Finns could also not view the nearby Germans as friendly. However, the Germans were farther away than the Soviets and hence less of a worry.

The Russian-manufactured 7.62mm Nagant was the first duty weapon of Finland before it adopted the 7.65mm Luger (known as their M23) in 1923. Why they chose it in 7.65mm rather than the more popular and powerful 9x19mm—especially with the experience of World War I behind them—is unclear to me. But the M23 served the Finnish military quite well until the 1980s, when it was replaced by the Lahti 9x19mm pistol that looks much the same

but has a much better gas system to help the pistol function in very low temperatures and also is in a better military caliber.

The comments about the Luger pistol in the section dealing with German handguns apply here as well. The M23 has the same standard Luger sight and safety system, yielding all the same problems. The 7.65mm cartridge has never been a successful military cartridge. The Swiss used it for years, but unlike the Finns, they never had to fight a war with it. The Finns used the M23 to fight two hard wars with the Soviet Union. I cannot imagine the 7.65mm was any better than 9x19mm, but perhaps it offered good penetration of the heavy winter clothing worn by the Soviet troopers.

In the test I conducted, the 7.65mm shot very flat, which would make hitting a distant target easier, but the combat pistol is designed for hitting close-range targets, not distant ones.

The trigger pull on the M23 tested is quite

nice. Unlike many Lugers, its pull is good and crisp. The muzzle blast with Fiocchi ammunition, which was supposed to be ex-Finnish army stock, was very heavy. Recoil was considerably lighter than that found in similar 9x19mm Lugers. The trigger guard on this Luger, as is common, is quite small. This must have caused a lot of problems in the Finnish climate, where gloved hands are standard.

Although this is an interesting pistol to inspect and shoot, I agree with the Finns that the M23 is not a suitable military pistol.

Author testing the Finnish M23 7.65mm Luger.

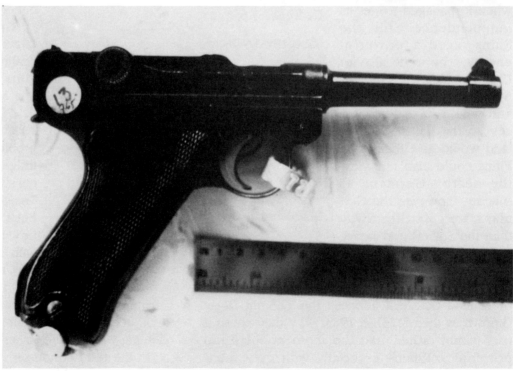

Right side of Finnish M23 Luger 7.65mm.

Close-up of the Finnish army acceptance mark (SA) on this M23.

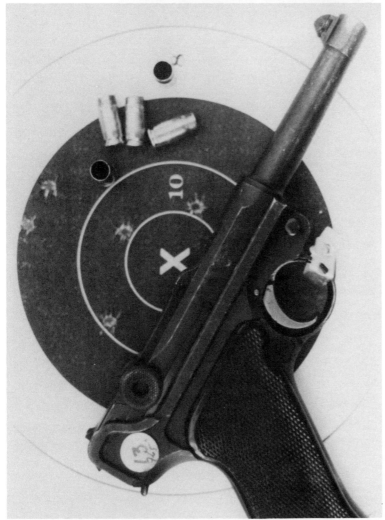

Finnish M23 Luger with 50-foot test target.

Lahti L-35

9x19mm
(9mm Para)

The Finnish Lahti tested was very similar to the Swedish M/40 Lahti and suffered the same shortcomings. It was quite heavy for its caliber, its magazines were difficult to load, and its sights no doubt would have produced the same poor results on the cinema range. As with the Swedish example, the weapon's grip was too sharp, and the safety was difficult to rapidly disengage. But where the Swedish Lahti was disappointing in its accuracy, the Finnish example shot good groups both with and without the shoulder stock. Also, trigger pull on the Finnish example was quite crisp. The Finnish Lahti tested was well broken in, so perhaps it seemed smoother in action because the Swedish example was basically unfired.

All Lahti pistols are rare, but if you frequent the northern parts of Europe, you are likely to find one. Similarly, since the Swedes have surplused off their Lahtis, they are not uncommon in the

SPECIFICATIONS

Name: Lahti L-35 9x19mm

Caliber: 9mm Parabellum

Weight: 2.7 lbs.

Length: 9.7 in.

Feed: 8-rd., detachable box mag.

Operation: Short recoil; semiauto

Sights: Front blade; rear notch

Muzzle velocity: 1,138 fps

Manufacturer: Valtion Kivääruthedas Jyväskylä

Status: Obsolete

United States. Although rarer than Lugers, Lahtis generally are cheaper, which I find odd since I prefer Lahtis. In Finland, the Luger M23 7.65mm was used prior to the adoption of the Lahti L-35, although many Lugers continued to soldier on. The Lahti L-35 was the classic Finnish Winter War weapon, and all reports praised its ability to function in the extreme climate of northern Finland. No doubt, however, it was an expensive weapon to manufacture.

Oddly enough, when the stock of Lahtis got low and Finnish armed forces needed a new weapon, rather than buying surplus Lahtis from the Swedes, the Finns bought Belgian P-35s. Perhaps there were problems with the L-35 that the Finns (who were the only people in the world to use it in combat) knew about that no one else did. Or perhaps they simply did not want to trust an old design and old metal for their soldiers.

The Lahti, whether the Finnish L-35 or

Swedish M/40, makes an interesting collectible. However, although clearly preferable to a Luger, it cannot be considered top of the line as far as 9mm handguns are concerned.

Author firing the Finnish L-35 Lahti with original stock attached.

Right side of Finnish L-35 9x19mm Lahti.

The 50-foot test target fired with L-35 with stock attached.

M1873 Ordonnance Revolver

11x17.8mmR (11mm French M1873)

The strangest part of testing this weapon was remembering how old it is, because many of its features are so modern that you tend to forget its age.

The pistol I tested was made in September 1880 ,but is in excellent shape. I fired a 1 15/16-inch group at 50 feet, demonstrating that accuracy is equal to a modern Smith & Wesson. When evaluating this weapon, you must keep in mind that it was a contemporary of the Colt Single-Action Army (SAA). When you realize that, it will suddenly dawn on you that this is a far superior combat weapon to the Colt. Except for the French ammunition going only 700 fps rather than the 900 fps of the Colt SAA, I would much prefer the French M73.

The double-action trigger system is like that of a modern Smith & Wesson revolver. The grip feels good and permits for fine instinctive shooting, allowing the trigger finger to correctly fall into position for rapid double-action work.

All the parts on the weapon are serial-numbered, and the whole piece shows great care in manufacture. As with all French weapons tested, the test target shows it to be very accurate. The markings and appearance all gave the weapon an elegant French appearance, which I find quite nice.

Reloading is slow because of the ejection system, but the double-action trigger is fast. I would prefer six 11mm rounds and their resulting slow reloading to seven 8mm Lebel rounds and fast reloading. The front sight has a McGivern-style bead that is quite useful for formal target work and quick acquisition on the cinema range. The light color of the front sight (due to absence of finish) allows for rapid pickup of the sight in poor light conditions.

This weapon was common in the French armies through World War I. Afterward, it was

SPECIFICATIONS

Name: M1873 Ordonnance Revolver

Caliber: 11mm

Weight: 2.7 lbs.

Length: 9 1/2 in.

Feed: Revolver

Operation: DA

Sights: Blade/notch

Muzzle velocity: NA

Manufacturer: State factory

Status: Obsolete

still in common service in the various French colonies. I know that if I had been forced to select a weapon for the trenches of World War I from among the M73, M92, or any of the other .32 ACP trash that the French ordnance people foisted on the French soldiers, I would have chosen the M73. The guys who designed this pistol knew what a combat revolver was for.

It is too light in caliber, but it is not impossible by any means. Perhaps the oddest thing about it is that it was issued without any finish. What the French did to keep them from rusting is beyond

me, but obviously it was effective. The example tested was almost 110 years old and still in excellent shape.

I dare say that when the test example was made, it was the finest combat revolver available, far superior (except for caliber) to Colt or Smith & Wesson revolvers available in 1873. The tested example had a 20-pound, double-action pull that was heavier than necessary, but perhaps it was intended to avoid mishaps in French cavalry units. It certainly did not greatly impede the performance on the cinema range. A good piece: I like it.

The author drew a very cold winter day to test the French M1873 Ordnance Revolver 11x17.8mm, a very good weapon.

The front sight on the M1873 was easy to pick up on a darkened range.

The rear sight of the M1873 is not ideal, but it is still better than many found on the same late nineteenth-century handguns.

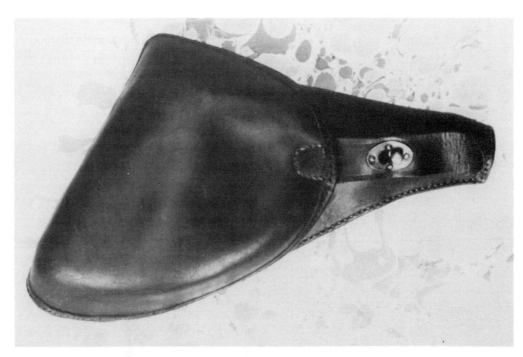

The holster used by the French army during the M1873's tenure.

The M1873 Ordonnance Revolver 11x17.8mm on 50-foot test target.

M1892 Lebel Ordonnance Revolver

8x27mmR
(8mm Lebel Revolver)

F was the first to adopt smokeless powder for its military. That decision was it was so important that it was a military secret, and it was a crime subject to severe punishment for a soldier to open a cartridge and display the propellant. For many who are unfamiliar with black-powder weapons, the importance of not having a large smoke cloud each time you fire may not be fully understood. A few shots with a black-powder weapon will quickly show the shooter how important this breakthrough actually was. Although standing behind the black-powder weapon will illustrate the importance, going down range and watching another individual fire will be even more revealing.

Not only is the lack of smoke an important element, but smokeless powder is also more powerful. It allows you to shoot projectiles at greater speed, giving you both more power and a flatter trajectory. Flatter-shooting weapons are always better, especially when considering weapons used at long range such as rifles and machine guns. Last, smokeless powder does not contain as much debris. You can fire a lot more rounds with smokeless than you can with black powder; smokeless will not foul the weapon.

After the French adopted smokeless powder, they developed the small-bore rifle to shoot it—the M1886 8mm Lebel rifle. This rifle represented a real breakthrough and was the envy of all other European armies. Unfortunately for the French, their technological lead was short-lived; the rifle's action was obsolete by 1891.

After the rifle, the French developed a handgun in the same bore to replace their M1879 11mm revolvers. Unfortunately, the French military had no way of knowing that the same principles that applied to rifles did not apply to handguns, but this soon became apparent. Although the 11mm

SPECIFICATIONS

Name: M1892 Lebel Ordonnance

Caliber: 8mm

Weight: 1.9 lbs.

Length: 9-1/4 in.

Feed: Revolver

Operation: DA

Sights: Blade/notch

Muzzle velocity: 715 fps

Manufacturer: State factory

Status: Obsolete

revolvers were adequate for stopping a charging native, the 8mm Lebel was not.

The pistol itself was well designed and well made of good materials. At a time when most solid-frame revolvers were rod-ejectors, the Lebel had simultaneous ejection. It broke to the right, as opposed to the left which is more common today, but this configuration is actually quite handy. If anything, you are less likely to hit your body with empty cases when ejected on the right, but the whole issue of left or right ejection seems inconsequential. The ejector rod has a large button on it, allowing you to eject the empty cases with a firm push of the hand, something that will hurt your hand with modern Colts or Smith & Wessons.

As might be expected, recoil is quite light and thus rapid double-action work is possible. The M1892 is a double-action revolver, and the double-action pull is quite smooth, if somewhat heavy with a little stacking. Single-action pull is crisp, and it aided me on the formal target field.

The front sight is quite high and has a bead formed into the front sight. That makes it somewhat difficult to hold on the formal target course (or perhaps just to my eyes, which are unfamiliar with bead front sights—although they were common with dueling pistols of the day), but they showed up quite well on the cinema range and helped rapid indexing. The rear sight is good also, and one tends to put the ball of the bead right on top of the wings of the rear sight, which is shaped like an Express rifle sight. This facilitates good acquisition and use of the sights in poor light.

The barrel is hexagonal in shape, which I think is quite elegant although perhaps somewhat labor intensive. The shape of the barrel also helps indexing, because the flat surface guided the eyes toward the front sight. Interestingly, the lightweight barrel

got quite hot after only 35 rapid rounds fired on the cinema range.

The grip feels good in my hand and does not need any adapter. Although the recoil is light, the grip would, I believe, still be good even with a more powerful cartridge.

Had this revolver been offered in 11mm but been converted to smokeless powder, this weapon would have made a wonderful combat revolver. Unfortunately, just as the French military would do again in 1935 when it adopted a nice pistol in a weak caliber, it doomed the Lebel from the start. Handguns are only bullet projectors, and thus they succeed or fail based on the cartridge. A lovely weapon in a poor caliber is never as good as a less cleverly designed weapon in a better caliber. Obviously, the goal is to have a well-designed weapon chambering a proper cartridge.

M1892 Lebel revolvers are found all over Africa and Asia, as well as in Europe, because of the French colonial and military presence. Familiarity with this pistol may come in handy when least expected. But do not forget that it is the cartridge that is important, so shot placement is critical since you are shooting what is basically a .32 Smith & Wesson long caliber, hardly a well-known stopper.

The light recoil of the 8mm M1892 Lebel is evident in this photograph.

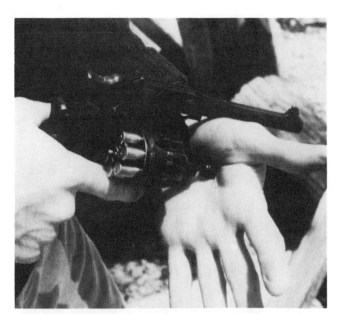

The large rod housing on the M1892/Lebel ejector knob allows the shooter to eject cases forcefully without hurting the hand. Ejection on the right avoids empty cases striking the shooter's thumb as sometimes occurs with right-handed shooters and left-sided ejections (above right).

The cylinder comes out on the right, which allows the shooter to reload quickly since the weapon is in the middle of the body (above).

The M1892 rear sight permits a definite sight picture (left).

The bead sight on the 1914 example I tested was helpful for formal target work, yet, because it was positioned high on the stem, it also provided good indexing ability in poor light.

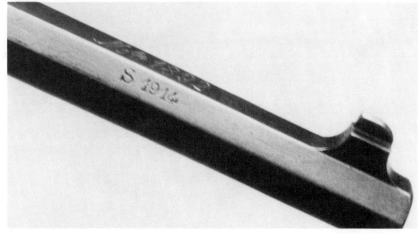

Pulling the latch to the rear causes the hammer to pull back, thereby preventing any hammer-primer contact.

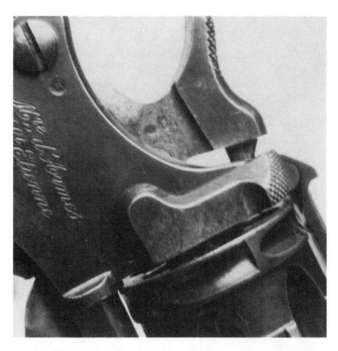

Returning the latch to a normal position allows hammer pin to hit primers.

Lebel M1892 8mm with 50-foot test target.

Ruby
7.62x17mmSR
(7.65mm Browning, .32 ACP)

This is another expedient French wartime pistol that then came into common police and civilian use after World War I. Apparently, a lot of these were taken home after the war; photographs in the German-controlled press during World War II showed weapons taken from members of the French underground, including these Ruby pistols.

Even today the weapons are common and inexpensive. Its being heavy and of standard design, I would be surprised to see any Ruby pistol ever wear out by its being fired. Rust, corrosive ammunition, or lack of care may reduce the Ruby to a hunk of worthless metal, but not actual use. As with many similar World War I autoloaders, the passage of time has now sufficiently obscured their original ownership, and lack of collector interest has kept them from being sold through normal channels for so long that you may well find this type of weapon anywhere in

SPECIFICATIONS	
Name:	Ruby
Caliber:	7.65mm Browning (.32 ACP)
Weight:	2.1 lbs.
Length:	5 7/8 in.
Feed:	In-line single column
Operation:	Blowback
Sights:	Blade/notch
Muzzle velocity:	NA
Manufacturer:	Gabilondo & Alkartasuna
Status:	Obsolete

the world. Hence, these old self-loaders of moderate quality have a real place in the world and should be respected for that reason.

This pistol is heavy (34 ounces) and has a fairly high magazine capacity (nine rounds) due its long grip. The safety is quite positive, but it seems to have a design glitch: you must shift the weapon in your hand to be able to pull it back and off, whereas it can be put on without shifting the weapon. Obviously this is backward, since it is more important to be able to remove a safety rapidly than it is to reengage it. There is no hold-open device when the last round is fired. If you apply the safety and insert the loaded magazine, pulling the slide to the rear, the slide will lock in place and not load. The safety becomes the slide stop when you pull on safe and the slide is pulled to the rear. This feature could get you seriously harmed because you cannot drop the slide and load the weapon until you shift the weapon in your hand.

The rear sight is pyramid-shaped, and that, coupled with the small, shallow front blade, makes indexing difficult. The rear sight also has points on it that can cause confusion with the front sight, especially when trying to use the weapon rapidly. The trigger guard is small, given the large trigger that fills it, and this could cause a real problem with a gloved hand. Although this might not be an issue for police weapons, this feature could be undesirable for military weapon that are frequently used in harsh climates.

There is not a lot to recommend this pistol: its sights are hard to use, its safety is hard to get off, and it is heavy and underpowered. I know I would prefer a M1874 French 11mm revolver to this "more modern" self-loader.

Author testing the Ruby 7.65mm.

Left view of Ruby 7.65mm.

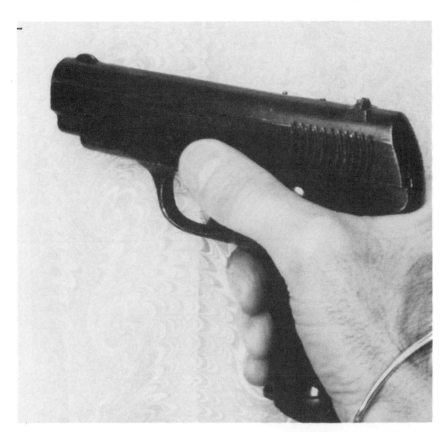

The placement of the safety requires that the shooter shift the weapon in his hand to disengage it (left).

The sights on the Ruby .32 ACP are very difficult to see and pick up easily (below).

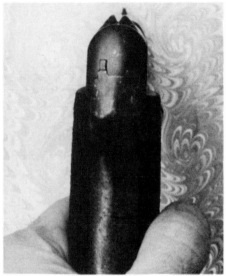

The Ruby .32 ACP with 50-foot test target.

Star "Military"
7.62x17mmSR
(7.65mm Browning, .32 ACP)

During World War I, the French military ballooned to a size that the planners never dreamed of. In addition to the fact that there were more people in the French military, more of the troops carried handguns than in any prior or subsequent war—a recipe for an acute shortage of handguns. This problem was not peculiar to the French; it was common to all nations during this period. The French responded in part by looking over the mountains to Spain and buying a large number of commonplace .32 ACP pistols to arm its troops. The Star Military was one of the designs selected.

Although the example I tested was obviously well worn and had not been taken care of as a prized collection piece, it shot a group of 2 5/8 inches, which is not bad. The rear sight was very small and shallow, and the front sight was also difficult to see—both of which made indexing on the cinema range quite difficult.

SPECIFICATIONS

Name: Star "Military"

Caliber: 7.65 Browning (.32 ACP)

Weight: 1.9 lbs.

Length: 7.5 in.

Feed: In-line, single-column box

Operation: Blowback

Sights: Front blade; rear notch

Muzzle velocity: NA

Manufacturer: Echeverris of Eibar

Status: Obsolete

The safety is much like that found on Smith & Wesson autoloaders, especially the single-action M52 and M745, except that it is worse because it's located on the wrong side of the slide. A right-handed shooter must break his grip to disengage it. Requiring the safety to be pushed up is also unnatural, at least to this old Colt Government Model shooter, though. I will acknowledge that you can train yourself to do it. Certainly, the French copied this safety system design on the M1935A, M19335B, and M50 pistols (although they at least shifted regarding the side of the slide), so perhaps the problems I would anticipate with the system never materialized.

The grip angle reminded me of the Colt Government Model, and hence instinctive shooting on the cinema range was enhanced. The lock of a hold-open device makes reloading difficult, but the nine-round magazine capacity minimizes this problem to an extent.

The straight-line feeding of the Star Military reminded me of the Heckler & Koch P7, which has this same feature and is known to be ultrareliable. I had no difficulty with the weapon during my test. The comfortable grip and the steel-frame pistol, which minimizes the recoil of the .32 ACP cartridge, make rapid repeat shots easy. Interestingly enough, although made to meet the critical demands of war, this pistol was better made, fitted, and blued than many modern pistols.

Considering its less desirable features, this pistol is heavy for the cartridge and has an awkward safety, and its sights hamper rapid indexing in reduced light. These features add up to a poor combat weapon, albeit a well-made one. Stars from this period are plentiful and have not attracted great collector interest, so they can be found all over the world when other more desirable pistols cannot.

Author testing the Star Military 7.65mm.

Right view of Star Military.

The 50-foot test target results of the Star Military.

Unique
7.62x17mmSR
(7.65mm Browning, .32 ACP)

The French military adopted this weapon as a substitute standard weapon. It is a heavy weapon capable of good accuracy with a 1 13/1- inch group fired at 50 feet. The open-hammer style, which allows you to check the weapon to see whether it is cocked or not, makes this a better weapon than the Ruby .32. Otherwise, the Unique and Ruby are quite similar.

The Unique is an all-steel weapon that weighs as much as or more than a Glock 17, SIG-Sauer P220 .45, but it is no way an equal to any of them. The Unique was designed after the Walther P38 was produced, and its obsolete design and poor caliber choice doomed it from the start. Still, the Unique is plentiful in many areas of the world, so familiarity is recommended.

The trigger on the example tested broke crisply, contributing to a good group. The wide rear sight allowed good pointing ability, although the narrow front made centering more difficult

SPECIFICATIONS

Name: Unique

Caliber: .32 ACP

Weight: 1.62 lbs.

Length: 5.7 in.

Feed: In-line box

Operation: Blowback

Sights: NA

Muzzle velocity: NA

Manufacturer: Unique

Status: Obsolete but still in use by police

than would otherwise be the case. The pistol lacks a hold-open device, so you do not know when the last round is fired—an obvious disadvantage. The safety is very difficult to disengage; you must shift the weapon in the hand to hit it, then return to the original position. Obviously, this slows up your reaction on target if the gun is carried cocked and locked.

The recoil is low because of the weapon's weight and its cartridge. The grip feels very similar to that of the Colt Government Model. Because of the small, dark, and narrow front sight, indexing on the cinema range was difficult, but, of course, this could be minimized by painting the sights. The smooth profile of the weapon does prevent snagging and tearing on the body. Having a nine-shot capacity is certainly an advantage, although it would take THV ammo to convince me that any .32 AP pistol is really suitable as a military weapon.

Author testing the Unique .32 ACP.

View of right side of Unique 7.65mm.

The butt-mounted magazine release slows reloading, but it does allow the left-handed soldier to work the weapon easily. Such systems do tend to enhance magazine reliability since you avoid the problem of magazine catches being slightly off. The Unique does have a magazine safety, which is desirable on a military gun. I have been present in a military police station when a .45 Colt Government Model was accidentally discharged. I have never seen anyone need to fire in the short time that the magazine was out and a new one was being inserted, nor have I seen anyone who has lost his magazine being reduced to using his weapon as a single loader. If you frequent the upper headwaters of the Amazon by yourself, perhaps this is a concern. If, however, you have to deal with low-skill, low-interest military people (officer or enlisted) who never fired a handgun before coming on duty, the magazine safety makes sense. The well-qualified people can use any type of weapon safety.

Overall, the Unique is a well-made, heavy weapon of good material but poor caliber and rather obsolete design. I would much rather have a Roth-Steyr 07.

Results of the Unique's 50-foot test target.

M1935A
7.62x17mmSR
(7.65mm French Long)

This pistol was developed by the French State Arsenals prior to World War II to replace the 7.65mm pocket pistols (imported from Spain during the World War I), the remaining 11mm Lebel revolvers, and the standard duty weapon of the period, the 8mm Lebel.

On the formal target range, the small front sight and light recoil yielded good groups, even using old French military ammunition, which is subject to hangfires and duds. I found that I was able to put five off-hand shots into 2 1/8-inch without any problems groups, which is smaller than what I shot with my standard Model 19 that day. I attribute these good groups to light recoil and the accurate sighting that resulted from the small front sights. The trigger pull in the example was heavy, and I really had to fight the trigger pull to shoot it properly. Its weight did enable me to pull almost all of it out and then concentrate on the last few pounds. I had no weapon-

SPECIFICATIONS

Name: M1935A

Caliber: 7.65mm long

Weight: 1.6 lbs.

Length: 7.6 in.

Feed: 8-rd., in-line, detachable box mag.

Operation: Recoil; semiauto

Sights: Front blade; rear rounded notch

Muzzle velocity: 1,132 fps

Manufacturer: NA

Status: Obsolete

related malfunctions, but the age and quality of the ammunition did result in some hangfires. The rear sight is too small and shallow, and the front sights are too narrow for quick indexing.

The weapon does feel good in my hand, and it is quickly broken down into individual components. Probably its most interesting feature is the hammer: it can be taken completely out of the weapon to detail-strip it. I am sure that the many people who served in French Indochina must have appreciated this feature when they detail-stripped their weapons after days in the swamps south of Saigon.

This weapon performed well on the target range but not on the cinema range, where the sights hindered the weapon's ability to index. If painted, however, the sights would be a lot better.

The trigger is curved in this particular pistol, and, as a consequence, your finger has a tendency to drag along the bottom of the trigger guard and thus bruise it during long firing strings. The safety requires cocking the thumb over the top of the slide to disengage it, which does require you to break your firing position. Still, it is better than attempting to cock the small hammer sheltered by the slide and frame and that uses a fairly stiff coil spring. To try to cock that under conditions of stress would be chancy. Even though it requires you to break your grip, this safety is better than that of many other European military weapons, such as the Browning P-35, and can be rapidly disengaged with practice. It is not as fast as the safety on the Colt Government Model but is better than many.

This is a very accurate, well-made, well-machined, well-designed weapon of top-quality material, with a safety that can be mastered. The worst thing about it is the caliber: .30-caliber, full-metal-jacketed, lightweight projectiles with low velocity are just not reliable when using conventional ammunition. This Petter design was purchased in 1937 by SIG, which basically designed its P210 around it. The SIG P210 is probably the finest 9mm pistol in the world, and it certainly can be appre-

ciated by anyone who likes fine handguns.

Overall, the M1935A has a reputation as a terrible handgun, but, other than its caliber, it isn't. And one should be familiar with these weapons because of their availability wherever the French had troops in the immediate postcolonial days, especially in Africa and Southeast Asia.

French civilian armed with M1935A in training during the Algerian War.

Author test-firing the M1935A.

The M1935A disassembled. The ability to break it down quickly makes maintenance easier.

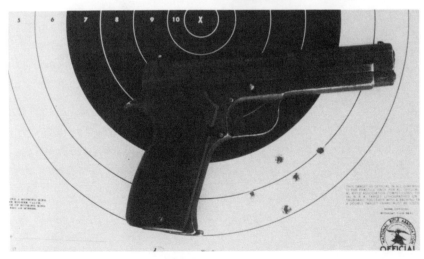

M1935A with the 50-foot test target.

M1935S
7.65x20mm
(7.65mm French Long)

For the most part the M1935S was used by the French in the hard-fought colonial wars of the 1945-60 period in Africa and Southeast Asia. In Algeria, in the 1950s, the M1935A & S models were issued to civilians who were at risk from the guerrillas, because the newly designed MAC 50 9x19 mm pistol was in short supply. Very few of the weapons were used in World War II because only a few had been made prior to the German occupation. About 40,000 were made during the war and procured by the Heereswaffenamt. Consequently, you are more likely to run into these weapons in Africa and Southeast Asia than in Europe.

As with all the French weapons tested, this one shot very good groups on the formal range. The ones that went outside the main group were pulled shots. Although this weapon is not nearly as elegant as the M1935A, it seems rugged. The finish was painted flat black, and I expect that it would be easier to maintain than the blued M1935A finish.

The safety is placed similarly to that of the M1935A, so the comments made about the latter's safety apply here as well.

Both weapons have an underpowered cartridge. Had the French manufactured this weapon in 9x19mm, it would in fact be better than a MAC 50 because it would be smaller and lighter, which is always better for military use. Failing that, they should have made them in .32 or .380 and avoided such an involved locking system. As it is, the design is elegant, except for the safety location. But the quality of design is lost with the odd 7.65 French long caliber that effectively precludes its use unless you have access to French army supply munitions.

SPECIFICATIONS

Name: M1935S

Caliber: 7.65mm Long

Weight: 1.8 lbs.

Length: 7.4 in.

Feed: 8-rd., in-line, detachable box mag.

Operation: Recoil, semiauto

Sights: Front blade; rear blade

Muzzle velocity: 1,132 fps

Manufacturer: NA

Status: Obsolete

Author test-firing M1935S 7.65mm (right).

Right side of M1935S (below).

Left side of M1935S (below right).

The 50-foot test target shot with M1935S (right).

MAC 50
9x19mm
(9mm Para)

Except for specialized units that use the MAB 15, the French army currently has the MAC 50 as its service pistol throughout the world. However, it is due to be replaced by the French-made M92 Beretta.

The MAC 50 was designed in the late 1940s and came into operation in the 1950s. In contrast to earlier French weapons—such as the 11mm and 8mm Lebel revolvers, 7.65mm Ruby automatic pistols that were common during the World War I and Model 1935A and 1935S pistols—the MAC 50 was certainly a big step in the right direction as far as service pistols are concerned.

The example I tested had a great trigger pull. I wonder if all French pistols have good trigger pulls, because all the ones I tested for this book do.

This pistol is fitted with a magazine safety, always a good feature on a military gun but sometimes deprecated by civilian shooters. The

SPECIFICATIONS

Name: MAC 50

Caliber: 9mm Parabellum

Weight: 1.8 lbs.

Length: 7.6 in.

Feed: 9 round in line

Operation: Recoil, semiauto

Sights: Blade/notch

Muzzle velocity: Standard for cartridge

Manufacturer: State factory

Status: Out of production; widely used in French military and police units.

safety on this weapon is somewhat unusual; it is located on the top of the slide, much like on the Model 1935A. Therefore, when pushing the safety off, you have to break your grip, unlike with the grip you take on the Colt Government Model safety. If you remember this while you are covering suspects, however, the safety can be disengaged quite rapidly. In fact, this safety is much faster to disengage than that on the Browning High Power P-35 pistols with a standard factory safety. But until it is fitted with the new extended safety, I prefer the P-35.

An interesting feature associated with this safety is how it attaches. If you carry this pistol in condition three—slide down, hammer down, chamber empty—and rapidly retract the slide by pulling it with your left hand while your right hand pushes on the receiver as you look toward the target, the safety is automatically engaged. Your finger seems to drag on it, and this causes it to slip on automatically as the slide advances. At first, I thought that I was not taking my hand off fast enough, but try as I would, it happened to me every time I tried to do it rapidly. If you do it slowly, you do not notice it because you get your hand off it properly or you run your hand down the side.

Leroy Thompson, who has worked frequently with French Foreign Legion troops, remembers this was as a commonly accepted technique in the Legion. In practice sessions with them, he always found that, when the slide was dragged to the rear to load, the pistol was handed to him with the safety already engaged. Until we tested this pistol, he had always assumed that the legionnaires put on the safety before handing the weapon to him. I don't know whether this safety design was intentional or coincidental, but it is a good feature.

The magazine release is well located; pushing the release with your thumb and withdrawing the magazine does not cause the slide to go forward, as is common with many military weapons designed in Europe. The slide stop is also well fitted, and with a simple touch the slide goes forward, loading the weapon if you have a fresh magazine in it or at least dropping the slide with the empty magazine. This weapon also has a loaded chamber indicator; the extractor pushes up when

the weapon is loaded. This is beneficial on military weapons because you can always see (or feel in the darkness) if the weapon is loaded.

The whole weapon looks a little odd (what I would call a typically French fashion), but it is, in fact, well made and of high quality. The grip ap-

French trooper with captured Viet Minh guerrilla during the Indochina War. Under the leadership of Giap, the Viet Minh evolved from a guerrilla army into the North Vietnamese Army. The French trooper is armed with the MAC 50.

pears rather long, but it feels good in my hand. Sights are adequate for the purpose of formal target work, but on the cinema range they were gray and low. This made it difficult to see or index well. However, a dab of white paint would

The MAC 50 being field-tested by the author.

certainly rectify this problem.

Even when it is in the down position, the hammer looks as if it is half-cocked. This does allow you to cock the weapon rapidly with the left, or weak, hand if the weapon is to be carried in condition two (not necessarily the most desirable carry method but commonly done in military circles). The weapon can be cocked rapidly from condition two (hammer-down position), but, as mentioned, the weapon's design also allows you to carry it in condition one (hammer cocked, safety on, chamber loaded).

Testing on the cinema range revealed one undesirable fact: the weapon had a lot of flash to it. Using the same ammunition I commonly use with other 9mm pistols, I found more flash with this pistol. Accuracy was not spectacular, but it was perfectly adequate with this weapon: five-shot groups fired within 4 1/4 inches.

Overall, the MAC 50 is much better than anything previously seen in French military services and certainly not a weapon to be discounted.

MAC 50 atop various French Foreign Legion material.

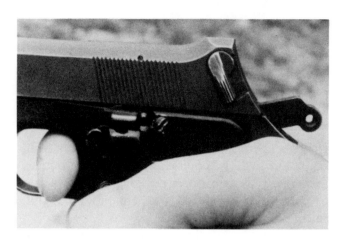

The MAC 50 with safety off.

The method used by author to disconnect safety when carried cocked and locked. It is not as fast as the method available on the Government Model M1911, but it is still a lot better than that found on Lugers.

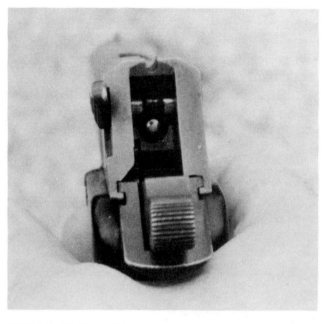

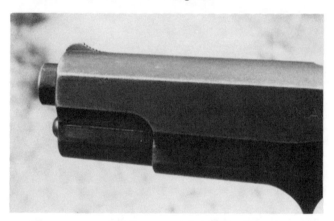

The front sight of a MAC 50 is low and too dark.

MAC 50 with the safety on.

MAB-15

9x19mm
(9mm Para)

The MAB-15 is a commercially available double-column, all-steel single-action pistol. According to the label on the box that accompanying the test weapon, it comes in nickel or blue finish and is available in .45 ACP caliber. I have never seen anything except the standard blue 9mm pistol myself; the others must be very rare, if they exist at all.

Certain elements of the French military and the paramilitary police organizations use this pistol instead of the MAC 50, M1935A, or M1935S pistols. Whether the adoption of the Beretta 92G model by the national police and, supposedly, the army will cause these weapons to disappear from the inventory is hard to tell. When in Paris in 1991, I saw military people carrying Unique .32s and MAC 50s. The elite police unit I was with had five members to the team; I saw no Berettas, but I did see a CZ 75, SIG P210, Smith & Wesson M59, and Colt Gov-

SPECIFICATIONS

Name: MAB-15 (MAB PA 15)

Caliber: 9mm Parabellum

Weight: 2.4 lbs.

Length: 7.9 in.

Feed: 15-rd., staggered-row, detach. box mag.

Operation: Recoil, semiauto

Sights: Front blade; rear adjustable notch

Muzzle velocity: NA

Manufacturer: MAB, Bayonne

Status: Current production

ernment Model .45. I would imagine the MAB-15 will be in the hands of various units well into the twenty-first century in France.

This pistol does not use a Browning-style locking system; rather, it uses a rotating-barrel method like the Steyr Hahn or Mexican Obregon. I fail to see why this is any better than a Browning lockup, and it would seem to require more machining on the barrel. The test weapon had a wonderful trigger pull. It was quick, short, and light. It was so light, in fact, that the first round fired was unexpected. This could cause problems in the hands of a poorly trained infantryman. The trigger guard is small, and putting a gloved finger through it is difficult, exacerbating the light pull in that situation.

Oddly, with the magazine removed, the slide cannot be withdrawn. This basically precludes you from removing the round in the chamber unless you completely unload the magazine first.

The Seecamp .25/.32 pistols have this same annoying feature. It is bad on a six-shot pocket pistol; it is worse on a 15-shot battle pistol. The safety is too small to flip on and off rapidly, and it is buried between the grips and the slide, which makes it difficult to flip out easily. The sights are hard to see due to being small, low, and dark. Painting them white would help a great deal. As they are, indexing in poor light was difficult.

The grip has a reverse taper to it; it gets wider at the bottom. This strains the bottom fingers by spreading them, thereby weakening the grip. As anyone can see by closing his hand and making a fist, the grip should be larger at the top if you want to maintain a hard, firm grip on something.

The weapons do recoil quite lightly because of the steel frame and the weight of a loaded 15-shot magazine. Removing the safety was slow and difficult, making condition-one carry unwise. The small, buried hammer is difficult to pull to the rear and cock rapidly; this makes condition-two carry slow and invites fumbling. Last, the magazine button stuck out quite a bit, and unless you take care in selecting a holster, you could very easily bump your magazine out.

About the only thing that recommends this weapon is its trigger pull. Otherwise, its sights, safety system, and grip are bad, and it is heavy. This weapon should be passed over if you can get a better autoloading pistol, such as the MAC 50 or M1873 revolver.

The MAB-15 being tested by the author. The rotary locking system seemed to cause twisting in the hand.

MAB-15 with 50-foot test target.

M1879
Ordonnanzrevolver
10.6x25mmR
(10.6mm German Revolver)

This weapon was adopted by Germany shortly after its unification in 1871 and used through World War I. There are even reports of its being used by third-string troops in World War II, which is not really so odd since U.S. Gen. George Patton packed a similar vintage Colt Single-Action Army (SAA) during World War II. Jeff Cooper, the noted firearms writer, said he also took a Colt SAA with him when he went to the Pacific in 1942. Basically, however, by 1914, this weapon was relegated to second-string troops in the German army; it was common in German East and West Africa during World War I.

In comparison with the Colt SAA, this weapon fares well except for power (the Colt threw a heavier bullet at considerably more speed). But, mechanically, the Ordonnanzrevolver was not so far off the mark. Of course if you compare it to the M1873 French 11mm revolver or the British

SPECIFICATIONS

Name: M1879 Ordonnanzerevolver

Caliber: 10.6mm

Weight: NA

Length: NA

Feed: Revolver

Operation: Single action

Sights: Blade/notch

Muzzle velocity: NA

Manufacturer: State factory arsenals

Status: Obsolete

Webley series that was coming on line at the same time, it pales considerably; either of those is a vastly superior fighting handgun.

The example tested was in excellent shape and showed considerable care in the manufacture. All the parts, including the grips, were carefully numbered to the weapon. The safety that was fitted was obviously useless, evidence of a garrison mind-set among the designers. Given the history of the German army around 1879, that makes sense. Safety is always more of a concern among garrison troops, whereas combat troops care more about efficiency.

The grips are quite good for instinctive shooting, and the low velocity of the ammunition (650 fps) gives the shooter the ability to fire good groups rapidly. The barleycorn front sight on the end of the long barrel allows quick indexing even with poor light on the cinema range. The low rear sight makes triangulation difficult if not impossible. The trigger on the pistol has a nice smooth finish and is sufficiently wide. It is exactly like the Smith & Wesson "combat triggers" now fitted to the fighting revolvers. The loading gate is very sharp on the edges and tends to cut the thumb. Reloading is slow because the cylinder has to be reloaded and the cylinder pin used to remove the empties. Obviously, sooner or later, it would be dropped in the mud, especially at night, and lost.

Based on my test target, sights are obviously regulated to 50 meters (54 yards), but the 2-inch group I shot satisfied me that the German trooper of 1880 was not at a terrible disadvantage vis-a-vis his American counterpart. I prefer an 11mm French M1873 to the German M1879. If, however, you find yourself stuck in some backwater or even in an advanced country like Britain (where the weapon's age and nonavailability of ammunition allow you to obtain it without a license), you could do a lot worse than the 10.6mm M1879. I think I would prefer it to almost any .32 ACP self-loader.

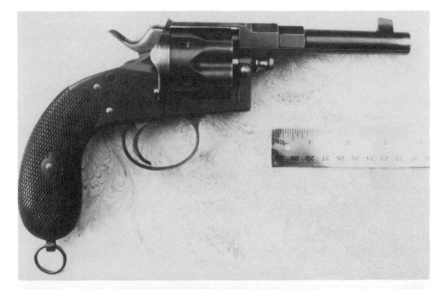

Right side of M1879 Ordonnanzrevolver 10.6x25mm.

The smooth combat-style trigger on this M1879 is very similar to that on the current Smith & Wesson revolvers.

A good view of the safety system of the M1879. Typical garrison thinking is evident here.

The barleycorn front sight hampered getting elevation correct on the formal range.

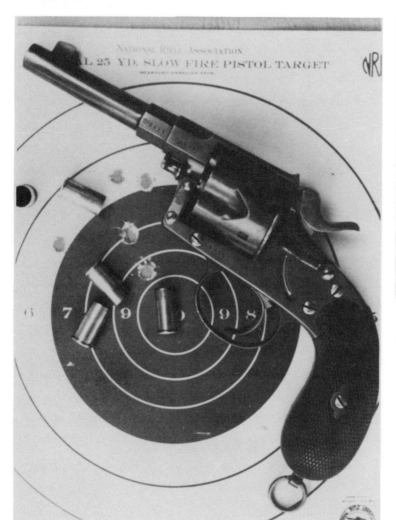

The front sight did allow good indexing in poor light, however (above).

Despite its age, the M1879 shot a good 50-foot test target (left).

Mauser M96

7.62x25mm
(7.63mm Mauser, .30 Mauser)

The Mauser model 1896 7.62mm was one of the first serious semiautomatic combat pistols in the world. The Mauser M96 or copies are still commonly encountered and respected in many of the more remote areas of the world. It would not be at all surprising to run across this weapon among drug warlords in the Golden Triangle of Burma or in places like Xinjiang in China.

The example I tested was once the property of a Chinese bandit, and thousands upon thousands of rounds had been fired through it. On the formal target range, accuracy was unspectacular, but interestingly enough, group sizes with the buttstock affixed were about half the size of those without it. The sights are very difficult to see. The rear sight is shallow (a narrow V shape), and the front sight is a large barleycorn, and this combination complicates the shooter's efforts to keep a proper target picture on the target range. Howev-

SPECIFICATIONS

Name: Mauser M96

Caliber: 7.62x25mm (Mauser 7.63; .30)

Weight: 2.7 lbs.

Length: App. 12 in.

Feed: Fixed box magazine

Operation: Short recoil

Sights: Front pyramid; rear notch or adj. slide

Muzzle velocity: NA

Manufacturer: Mauser

Status: Obsolete

er, you have to remember that this pistol was not designed for the formal target range. Although the sights are calibrated to some 1,000 meters, this is really not a 1,000-meter pistol by any means. If it were in good condition and had the buttstock affixed, however, it might be a serious 300-meter pistol. It could be viewed more as a carbine than a pistol under those circumstances.

Where this pistol really came into its own was on the cinema range. I have shot Mauser military pistols before, but I am like many people: I fired them on casual or formal target ranges and never really fired them under combat conditions. Therefore, I always viewed them as obsolete. So when I took it to the cinema range, I expected the weapon to exhibit violent recoil and twisting in my hand—basically, that it would prove to be a very ineffective combat weapon. This is *wrong!* Winston Churchill was certainly wise to carry one of these in his Light Horse regiment in the 1899 Boer War. I was quite shocked, to tell you the truth, at how well this pistol performed on the cinema range.

I commonly fire roughly 400 rounds per week through a Colt Government Model on the cinema range. Thus, I am quite familiar with how cinema ranges operate and how the Colt Government Model feels in my hand. I found my performance with the Mauser pistol, firing without the buttstock of course, to be as good as my performance with the Colt Government Model. The safety engaged and disengaged rapidly, and my thumb held the safety in position and flipped it off readily when covering suspects on the cinema range. First shot and repeat shots were also quick and on target. The high front sights on this weapon aided rapid and good indexing, which was clearly apparent in the poor light conditions of the cinema range. The grip did not shift in my hand as I had anticipated. The barrel did get hot after only 30 rounds, but that is not surprising in light of the small diameter.

This pistol is bulky for the caliber, and it is too complicated to be made cheaply enough for modern militaries. It is still in production in China, however, where it is marketed in selectable-fire mode with detachable magazine stock and bayonet as the Type 80. But if you are going into a situation in which you have a Mauser military pistol loaded with factory-fresh 7.62 ammunition, you are by no means at a disadvantage. This is a fine, well-designed combat—*not target*—weapon. In fact, with the general adoption of ballistic vests, it may even be a good caliber for that situation. The high-velocity .30-caliber loads are far in excess of the 1,600 fps of standard military loads. One could only wonder what would happen if you used loads such as THV in .30 Mauser. You might very well find that your loads would exceed 2,600 fps, because THV .357 Magnum loads hit that velocity level.

The Chinese bandit who formerly had owned this particular pistol was superbly armed for the challenges that his profession thrust upon him. If you find yourself having to choose between this weapon and a newer pistol or a heavy-frame revolver, you should give serious consideration to the Mauser. It may look obsolete, but it is in fact a good fighting handgun. I recommend it highly.

The Mauser M96 was very popular in Asia.

Here, the author fires the Mauser "Red Nine" with stock affixed. The weapon was previously owned by a Chinese bandit.

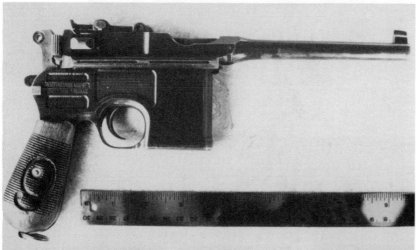

Right view of M96 "Red Nine" 9x19mm.

The poor performance of this M96 with stock resulted from a nearly shot-out bore caused by too many years of corrosive ammunition in the Orient.

Luger P.08
9x19mm (9mm Para)

One of the most familiar silhouettes in the world is that of the P.08 Luger. It was a much-favored souvenir of two world wars, and to this day has a large following that both likes and uses it, largely because it is a very accurate weapon.

Chic Gaylord, the famous 1950s New York City holster maker, told a wonderful story of a criminal holding a large number of New York City detectives at bay with a Luger P.08. They had .38 snub-nosed revolvers and shotguns and were unable to hit him. He kept their heads down from 75 yards away by smashing concrete scant inches over their heads. The fact that the barrel is fixed to the frame helps with accuracy.

The design of the gun makes the action very strong. Triggers are often quite bad, but an occasional good example does pop up. Sights are the typical European barleycorn front and V-shaped rear. That helps some in indexing on the dark-

SPECIFICATIONS

Name: Luger P.08

Caliber: 9mm Parabellum

Weight: 1.9 lbs.

Length: 8.75 in.

Feed: 8-rd., in-line, detachable box mag.

Operation: Recoil, semiauto

Sights: Front blade; rear V notch

Muzzle velocity: 1,050 fps

Manufacturer: NA

Status: Limited standard issue in Norway

ened cinema range because the front sight on the skinny barrel picks up fast, but the dark, shallow, low, and difficult-to-see rear sight slows your ability to triangulate on the range. The grip is quite good and is copied today in the Heckler & Koch P7 and Glock 17 pistols. They feel good in your hand, just like the P.08.

The pistol has a loaded-chamber indicator, but it does not have a magazine safety. The standard P.08 does not have a grip safety; thus the shooter must depend on the side safety only, which is poorly situated for quick removal and reapplication. You have to break your firing grip to remove it or apply it. I would not trust a striker-fired autoloader in condition one and certainly would not carry in condition zero. Because of the P.08's design, you cannot carry it in condition two since there is no way to safely release the firing pin. You are almost forced to carry the weapon in condition three, which is not fast enough to get into action

Right view of P.08
Luger 9x19mm.

for my tastes and requires the use of both hands. One of the main reasons to have a handgun is to allow you to defend yourself when you have only one hand available.

Many people have reported over the years that Lugers are unreliable, but I have never encountered any problems with functioning. Most problems that people report are traceable to poor magazines or low-power ammunition. Assuming that you have a pistol with matching numbers, you should use ammunition that is rated to what we in the United States call +P power levels, as well as good magazines. There are a lot of mismatched clunkers about, which may well have functioning problems.

The action is open at the time of firing, allowing mud to get into the action. At one time, people used to think this was a cause of malfunctions, but I do not believe so. The problem is the closely fitted parts, which are common in many German handguns. When they get mud in them,

the parts lose their ability to move, and the Luger is full of such parts. The German solution at the time of the pistol's introduction was to use a totally enclosed holster. Such a rig is too slow for a combat handgun, but apparently was necessary in the mud of France to keep the weapon clean.

Lugers have been popular since they first came out. You can find commercial and military models all over the world; it has been manufactured in four countries in innumerable variations. Knowing how to work a Luger, and what its strengths and weaknesses are, can really come in handy. Although I think a Luger pistol is one of the worst semiautomatic pistols for military purposes, many people like the mystique that goes with them, and they are generally very accurate. If offered one, you should take it, get powerful ammo for it, and then try to swap it for something better as quickly as possible. Given its mystique, you are likely to find someone who will swap you a useful handgun for your Luger in no time.

Left view of P.08
Luger 9x19mm.

P.08 Luger 50-foot
test target.

"Artillery" Luger
M1917-P.08 Lange
9x19mm
(9mm Para)

The Artillery Luger was designed during World War I to equip troops who could not carry a rifle but who needed an accurate weapon that provided more sustained fire capability than a traditional handgun. German machine gun troops were commonly equipped with this weapon, and it was frequently used with a 32-shot drum magazine. The 6-inch barrel length of the navy model no doubt was familiar to one and all because it was adopted before the army adopted the P.08.

When testing the weapon, I fired it both with and without the stock. I discovered that with the Luger, as with most other stocked pistols tested, I did not get smaller groups with the stock when I used a two-hand Weaver stance. Instead, I got the same size groups with the stock, but the firing was much easier. Additionally, I tested this weapon when I was running, crawling, and crossing distances, and I found that the stock

SPECIFICATIONS

Name: "Artillery" Luger (Model 1914 or 1917)

Caliber: 9mm Parabellum

Weight: 2.1 lbs.

Length: 12 3/4 in.

Feed: In-line single column

Operation: Short recoil, rising toggle joint

Sights: Front blade, adjustable rear notch

Muzzle velocity: NA

Manufacturer: DWM

Status: Obsolete

made it easier to hit targets when I was out of breath and my chest was heaving. No doubt, German machine gun crews during World War I, who were crawling in the mud and dodging artillery shells, found the same thing. Last, keep in mind when evaluating results that the normal way for shooting handguns during World War I was one-handed, and the stock basically forces the shooter to adopt a Weaver-type stance. So I conclude from all this that a stocked pistol has some valid use when issued to the military man whose normal duties would preclude his carrying a battle rifle. This was certainly more true in 1914, when the Mauser rifles had a 29-inch barrel, than today; but even now I can see that a lightweight butt stock would come in handy for a helicopter pilot, artilleryman, or combat engineer.

The Artillery Luger suffers from all the same drawbacks as does the common 4-inch-barrel model. The sights are hard to see on both the conventional and cinema ranges. The safety is difficult to disengage rapidly, and the trigger is not very good. However, the weapon does shoot quite adequately. At 300 yards, I found it easy to put the shots out fast and to engage 12-inch plates with it.

All in all, I would say this stocked Luger was a good concept, but its execution leaves something to be desired. I rather fancy the concept of the Stechkin 9mm pistol or Heckler & Koch VP70 pistol-stock combo, but a lack of weapons to test has limited my research on the subject.

The author test-firing the Artillery Luger with and without stock (above and right).

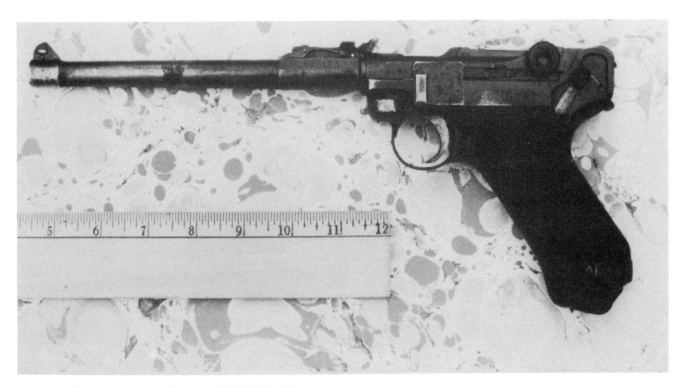

Left view of the Artillery Model Luger M1908L 9x19mm.

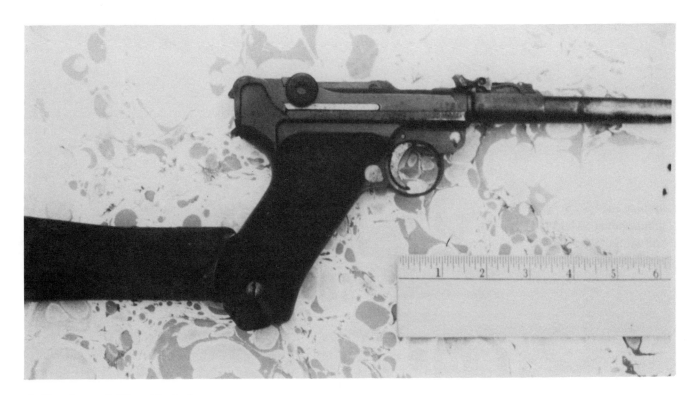

Artillery Luger P.08L with stock.

Artillery Luger P.08L with group shot at 50 feet with and without stock (above and right).

Dreyse

7.62x17mmSR
(7.65 Browning, .32 ACP)

More than 250,000 of these particular pistols were made prior to the end of World War I; 80,000 of them went to the German military, and the rest were apparently sold to private individuals who carried them during the war.

The Dreyse is an inferior pistol, especially when you consider it has the same power-to-weight ratio as the Model 03 Browning, which is a much better combat gun. Actually, I would go so far as to say that the Model 94 Nambu 8mm pistol, which is commonly deplored by most people, is a much better pistol than this one.

The Dreyse has only one good feature: the quality of the machining is superb, which makes stripping a breeze. Clearly, the machining skill of the people who made this particular weapon was very high. It is, however, a typical German gun with too many parts (all of them small) and too many springs. German gun designers just seem to

SPECIFICATIONS

Name: Dreyse

Caliber: 7.65 Browning (.32 ACP)

Weight: 1.5 lbs.

Length: 6 1/4 in.

Feed: In-line, single-column box

Operation: Blowback

Sights: Front blade; rear notch

Muzzle velocity: Standard for cartridge

Manufacturer: Rheinishche Metallwaren & Maschinenfabrik

Status: Obsolete

love using three parts to the job of one.

The sights on this particular weapon are terrible. They are down in a dished-out groove in the top of the slide, and there is a tendency (particularly in fast combat-range situations) to pick up the side of the slide as the sight rather than the sight itself, causing you to be way off mark at anything other than point-blank range. Also, the serrations on the side of the slide are sharp, and the spring is heavy. This, coupled with the sharp serrations, means that by the end of my test, my thumb was shredded.

Group sizes with the Dreyse pistol fired at the 50-foot formal target range ran 3 7/8 inches. This is not bad, considering the sights and length of the barrel, but it was on the cinema range that its deficiencies as a combat weapon were quickly obvious.

The safety is located in an awkward position. With practice, you can flip it off with one hand, but it is slow and uncertain. You really cannot reengage it without using two hands. Needless to say, you may not always have two hands to reengage a safety. The thought of taking a cocked, loaded Dreyse pistol and shoving it back into my belt without putting on the safety is enough to make my blood run cold.

It has other design flaws as well. It has a poorly designed trigger pull, because you pull through it to withdraw the firing pin and then it breaks very suddenly. It has a rounded grip that is sharp at the same time, which causes the

weapon to shift in your hand. It kicks considerably more than it ought to for the little 7.65 cartridge it shoots.

Handguns similar to the Dreyse, but chambered for the 9mm Parabellum, cartridge were common during World War I, and one can only suppose that those weapons with a more potent cartridge would be worse than this one.

All together, the Dreyse is an unsatisfactory automatic pistol: it is hard to use, not very safe, and low powered. It is possibly one of the worst tested.

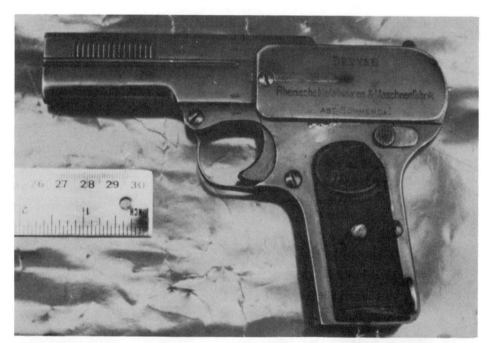

Left view of Dreyse 7.65mm pistol.

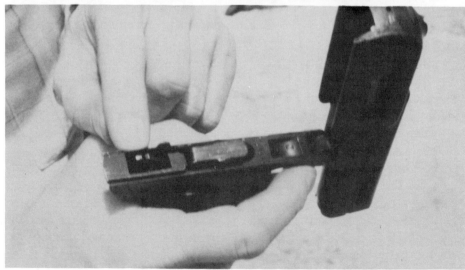

The author points to the hammer on the Dreyse. Light hammer blows are a problem with this design.

This photo illustrates the weapon with action open. Malfunctions are common with this weapon.

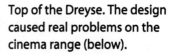

The awkward positioning of the safety made it difficult to apply and disengage (left).

Top of the Dreyse. The design caused real problems on the cinema range (below).

The 50-foot test target for the Dreyse.

Sauer Model 30 "Behorden"

7.62x17mmSR
(7.65mm Browning, .32 ACP)

This is one of myriad examples of .32 ACP self-loaders used as substitute handguns by the Germans during World War II that now, as a result of the war and its aftermath, turn up all over the world with little or no past history associated with them. And that anonymity, in large part, is their claim to fame.

However, the Sauer Behorden is a step above many .32 pistols pressed into World War II service. First, it is quite accurate. One of my test groups was only 1 1/4 inch in diameter even though the day was alternately cloudy and shiny, complicating my efforts to keep the sight accurately lined up. This in itself is remarkable for such a small pistol with so a short radius, but there is more. It is small enough to slip into a pants pocket and not be seen again until withdrawn. It will even fit in Levi's pockets, but, obviously, the front pocket of a German army uniform would be better.

SPECIFICATIONS

Name: Sauer Behorden Model 30

Caliber: 7.65mm Browning; .32 ACP

Weight: 1.4 lbs.

Length: 5.79 in.

Feed: Single line box

Operation: Blowback

Sights: Front blade; rear notch

Muzzle velocity: NA

Manufacturer: Sauer

Status: Obsolete

The safety is located so that you can flip it off and fire it quickly. The rear sight is high, and, although narrow and dark, it allowed for quick indexing on the cinema range. In addition to the normal safety, this pistol has a safety on the face of the trigger, as can be found on the Glock 17—another nice feature on a pistol likely to be carried in the front pants pocket. The firing indicator on the rear allows quick touch verification that the weapon is loaded and ready to fire. The magazine safety is fitted, which is always useful on any military pistol.

Recoil is brisk due to weight and size, but the design of the grip prevents any type of damage to the hand like that which commonly occurs with weapons such as the Walther PP/PPK. This pistol is also available with an aluminum slide and receiver and weighs 15 ounces. I wish I could locate one!

Although the caliber is weak for a military weapon, it is still an excellent choice for a military handgun because you could carry it in your pocket, and a rifle carried in the normal fashion would not interfere with it. Among the .32 auto that I have tested, I believe this is the best. In fact, I think it may be as good as the Modern .32 DA Seecamp, which has been my constant companion in my left front pocket, backing up my .45 or .357 Magnum, since 1990.

The German soldier who had the good luck to draw this pistol as his substitute handgun was indeed a lucky man. While not as common as other types nor as well regarded by the less knowledgeable, it is my first choice of pre-1945 .32 autoloaders, with the Sauer M38 being second. If you see one, buy it!

Sauer Model 30 with 50-foot test target.

The Sauer Model 30 being tested by the author.

Ortgies

7.65x17mmSR
(7.65mm Browning, .32 ACP)

This pistol was in commercial production in Germany and was also used as a substitute standard during World War II. As noted earlier, it was apparently a common practice to issue a small-caliber pistol and 25 rounds to each German soldier living in occupied areas for his personal protection. Certainly that approach makes sense, and it also explains the wide variety of small-caliber pocket pistols used by the German military during World War II.

The sights on this pistol are very hard to see, and the rear sight is quite shallow. While I was able to achieve a 2 5/8-inch group with the weapon, it took a lot of effort to do so because of the poor sighting system. On the cinema range, it was worse.

The smooth outline of the weapon makes carrying it in your pocket quite handy. However, what is really good about this otherwise commonplace steel-framed .32 ACP autoloader is the

SPECIFICATIONS

Name: Ortgies

Caliber: 7.65mm (.32 ACP)

Weight: 1.4 lbs.

Length: 6.5 in.

Feed: In-line, single-column magazine

Operation: Blowback

Sights: Blade/notch

Muzzle velocity: Standard for cartridge

Manufacturer: Deutsche-Werke

Status: Obsolete

safety system. The weapon has a grip safety that must be pushed in to disengage. To put it back on safety, you push the button under the thumb, and the grip safety is then reapplied. It is very quick to reapply and disengage. The weapon can be loaded without disengaging the safety, thereby avoiding the "hot gun" problem. You can also cover a person with your weapon without disengaging the safety because merely closing your hand automatically disengages it, safety allowing instant fire. If you pull the weapon out of your pocket or holster by the front strap (not push in on the rear), the safety is still engaged, preventing the weapon from firing. This approach is much better than that found on the Heckler & Koch P7, where the safety is on the frame front and hence automatically disengaged when drawing the weapon.

The caliber is too light, but this is otherwise quite a good pistol with its smooth, snag-free finish and fine safety system that is designed to allow the weapon to be carried loaded with a round in the chamber, safety on. I prefer it to the more common Walther PP/PPK series for these reasons.

Because of the popularity of these weapons among civilians and their obvious nonmilitary caliber and appearance, these weapons can turn up in many odd corners of the world. Familiarity with the design could prove invaluable if you were to get stuck someplace where a 60-year-old pistol made by a company long out of business would look like a long-lost friend.

Right view of Ortgies 7.65mm.

Left view of Ortgies 7.65mm.

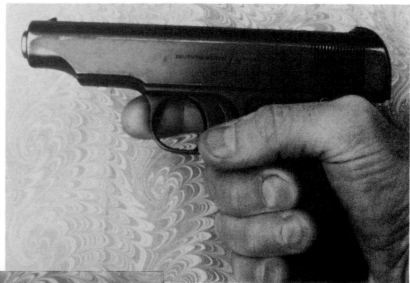

By pushing in with the hand, the shooter can quickly remove the safety and fire the weapon.

To reapply the safety, simply open your hand and push in with your thumb.

The 50-foot test target for the Ortgies.

Mauser HSc

7.62x17mmSR
(7.65mm Browning, .32 ACP)

This pistol is another of the pre-World War II commercial pocket pistols designed in 1930s Germany. It was a direct competitor to the Walther PP series, offering similar trigger systems, caliber, weights, and quality of manufacture. During the World War II, thousands of these weapons were used as a substitute pistol in the German armed forces, and, as a result, they are widespread today. After the war, they were manufactured for a time but are no longer available from the manufacturer.

For some reason, this pistol was never as popular as the Walther PP/PPK series, though it is really hard to see why. The slide on the Mauser does not cut your hand the way the Walther does, and that is a major improvement. All too frequently, I have had my hands cut when shooting Walthers.

The Mauser HSc sights are hard to see, making

SPECIFICATIONS

Name: Mauser HSc

Caliber: 7.65 Browning (.32 ACP)

Weight: 1.3 lbs.

Length: 6.5 in.

Feed: 8-rd., in-line, detachable box mag.

Operation: Blowback, semiauto

Sights: Front blade; rear round notch

Muzzle velocity: 950 fps

Manufacturer: Mauser-Werke Oberndorf

Status: Under license to Armi Renato

rapid indexing on the cinema range difficult. Both front and back are low and small, but painting them white would be a quick fix. On the formal range, the small size and narrow notch made it difficult to hold elevation. The pistol seemed to shoot fine: I put three shots into 1 1/2 inches, but because of the difficulty of utilizing the sights, my other two shots took the group out to 4 1/2 inches.

The trigger has a funny little hook at the bottom that felt odd in my hand and cramped my finger. It would be better if the trigger curve was less pronounced. The magazine release is butt-mounted, which, though slow, does prevent the holster from bumping out the magazine (but car seats *can* do this, so watch out). The magazine has an extension on it, like those found on Walther PP/PPK pistols, that is designed to allow the larger hand to hold the pistol better. Unfortunately, the extension is designed backward, because it stretches the little finger. It should be 180 degrees in the other direction, like the extension found on the ASP conversion of the Smith & Wesson M39. The little finger would then have a place to hang on a big hand and not be stretched out of place.

The pistol does offer a hold-open device indicating when the last round has been fired, a desirable feature commonly missing on many self-loaders of this vintage.

The safety works like any slide-mounted safety except that you must take care to push it off firmly. Many slide-mounted safeties are on only when firmly put in place and are off when slightly ajar. This is not so with the Mauser HSc. Only when the safety is definitely pushed back in the off position will the weapon fire. This feature is, I suppose, desirable, but one you must be careful to note. A couple of times when testing the weapon, I failed to get the safety all the way off, and the weapon did not fire. Further, unlike the Walther P38, when the safety is not all the way off, you cannot simply click the weapon and get it to go off. I found that with the Walther P38, if the safety was not totally off, simply pulling the trigger a couple of times would knock the safety off and allow the weapon to fire because the hammer fall would rotate the safety into the fire position. This is not so with the Mauser HSc.

The double-action trigger is heavy and thus hard to control when firing, but it would be safer to carry it with the safety disengaged.

The tested example was in .32 ACP, but these weapons are also found in .380 ACP. I prefer the .380 version, but there really may not be as much difference between them as we Americans like to think. I have heard of this weapon in .22 rimfire, but I have never seen one. At 20.5 ounces, the Mauser

The author firing the Mauser HSc 7.65mm.

The 50-foot test target for the Mauser HSc.

is too heavy for its caliber, in light of lighter-weight pistols like the Glock 17. Still a Mauser HSc is a good pistol and is sufficiently plentiful around the world so that everyone should be familiar with it.

Sauer Model 38

7.62x17mmSR
(7.65mm Browning, .32 ACP)

The German army also used the Sauer Model 38 pistol as a substitute weapon during World War II. It was designed just before the start of the war, and a variety of examples exist: some with and some without the external safeties. The model I tested was a wartime production gun in fair condition with no external safety.

The Sauer has a concealed hammer, which means you do not have to worry about dust and other crud coating it. It uses a double-action trigger system. It has a decocking lever, which is nice because you can cock the weapon and then drop the hammer to allow it to fire on double action only. Subsequent shots can be fired in single action, but at any time you can drop the hammer and start out with double action only, keeping it safe. Unlike the Walther PP/PPK series of pistols that I tested, I encountered no slide or hammer-bite problems with the Sauer. This is a better pis-

---

SPECIFICATIONS

Name: Sauer Model 38

Caliber: 7.65mm

Weight: 1.3 lbs.

Length: 6.25 in.

Feed: In-line, single-column box

Operation: Blowback

Sights: NA

Muzzle velocity: NA

Manufacturer: Sauer

Status: Obsolete

---

tol than the Walther series because its concealed hammer makes it easier to get out of the pocket and it does have a loaded-chamber indicator.

Groups from this pistol from the off hand were four shots into 2 inches and one flyer that took it out. With its sights and shorter length, however, I do not consider it too bad a performance by any means. Group size was close to that of a Model 19 group, if the flyer is excluded. The pistol also performed well on the cinema range, limited only by its caliber.

It has a good safety, a decocking lever, and it felt good in my hand. The sights are small but allow good indexing. If painted white they would be faster, but even black they were not bad. The sealed action protects against dust and dirt getting into the action. Of the .32 ACP pocket pistols that were tested, I would rate this number one overall. It has a good trigger, sights, indexing on the cinema range, and accuracy; is safe to carry; and has no hammer bits.

If I could get this pistol in .380 (some were made in .380 for military purposes), it would be my favorite .380 pocket pistol. It might not be equal in power to a Smith 649, for instance, with .38 Special 158-grain hollow points, but it would not be bad by any means.

Caliber, of course, is always a serious issue with these pistols. Burst-fire methodology (whereby a magazine empties into the intended target) is the only way to offset this problem, and you are basically converting the pistols into single-shot shotguns. Even so, they were lacking in stopping power. However, with the adoption of such loads as the French THV .32 ACP round, this deficiency might be lessened somewhat.

If you have the choice between a Sauer Model 38 or a Walther PP/PPK, pick the Sauer.

Left view of Sauer M38 7.65mm.

Detail of the rear of the slide, illustrating the Sauer M38's hammerless design.

Close-up of the decocking lever of the Sauer M38. This feature is now commonly copied on SIG pistols.

Sauer M38 7.65mm on 50-foot test target.

Mauser 1910/34

7.62x17mmSR
(7.65mm Browning, .32 ACP)

This pistol is well-made, but its design is terrible. Although the heel-butt-mounted magazine does not close the slide when withdrawn as with many European pocket pistols, it does close when it is reloaded. I assume this is designed to load a cartridge upon closing, but the heel-mounted release does make it slow to reload.

The example tested had a number of malfunctions, but I put this down to a faulty magazine. When the magazine is out, the weapon will not fire, and when the safety is applied, you cannot withdraw the slide. If you choose to carry it with the chamber empty and magazine loaded, as you might because of the safety problems with a striker-fired weapon, you must take care not to apply the safety. If you do so, you will not be able to chamber a round until it is disengaged, and this yields the worst of all worlds: you must push a small safety button to disengage before you can even pull the slide back to

SPECIFICATIONS

Name: Mauser M1910/34

Caliber: 7.65mm

Weight: 1.3 lbs.

Length: 6.2 in.

Feed: In-line box

Operation: Blowback

Sights: Front barleycorn; rear notch

Muzzle velocity: NA

Manufacturer: Mauser

Status: Obsolete

load, and you have a disengaged safety when it is loaded, which is similar to that found in the CZ 27. You must push directly on the button to disengage the safety since it is shrouded by the wood strip. A number of times on the cinema range, I found that the thumb pressure I exerted was insufficient to disengage the safety, and that I had to shift the weapon in my hand so that my bony knuckle would push on the safety button. Since the button does not protrude much past the grip, I could not put enough pressure on it to disengage, and the weapon would not fire. This is a dangerous feature, and unless you are careful in your holster selection, it could be even more so: the safety could easily be disengaged during withdrawal. This system was great on the formal range, but on the cinema range, it was slow and uncertain—exactly the same problem with the CZ 27.

Deciding how to carry this pistol is complicated by the safety problem. The front sight is a

European barleycorn design, small, and very dark, which greatly inhibits indexing. The rear sight suffers from the same problems. The grip feels comfortable, except that the magazine has a projecting spur that hit my small finger. Thus, when firing, I attempted to shift my grip to avoid the spur, thus slowing down subsequent shots. I would grind off the spur, but it is needed to retrieve the empty magazine because the heel-mounted magazine release demands something to pull on when the empty magazine is withdrawn.

The trigger has a considerable amount of slack before it fires, but it can be pulled through on the formal range and really does not cause any problems on the cinema range.

This is an inefficient design, and one can see why the design of the HSc or Walther PP/PPK was so welcomed by pocket-pistol carriers. The Mauser 1910 is not the worst .32 ACP tested, but it is very close.

The author testing the M1910/34. Note the empty case leaving the weapon.

The M1910/34 and the 50-foot test target shot.

Walther PP/PPK
7.62x17mmSR/9x17mm
(7.65mm Browning, .32 ACP/.380 ACP)

Walther PP/PPK pistols are world famous. This notoriety is due more to Ian Fleming's arming his fictional hero James Bond with one, I believe, than any real value of the pistol. Although not the first series to offer a double-action trigger style (Czech pistols had it first), the Walthers had that feature and were available in large quantity, and they were first to offer a production 9x19mm pistol with such a trigger system.

The Germans used Walther PP/PPK pistols as substitute pistols during World War II, mainly in .32 ACP, occasionally in .380, and even more rarely in .25 ACP. A few are seen in .22, but I do not believe they were ever meant to be military weapons (although many people view a Walther PPK in .22 with a sound suppressor as a fine military weapon).

These pistols were available commercially before World War II around the world, and after

SPECIFICATIONS
Name: Walther PP/PPk
Caliber: 7.65mm (.32 ACP, .38 ACP)
Weight: 1.5 lbs.
Length: 6.8 in.
Feed: 8-rd., detachable box magazine
Operation: Blowback
Sights: Front blade; rear notch
Muzzle velocity: 943 fps
Manufacturer: Walther Waffenfabrik
Status: Obsolete

a few years of their absence following the war, models made first in France and later in Germany reappeared on the world market. Additionally, copies of this design are made in a variety of countries. I have seen an exact 1950s copy made in the People's Republic of China, no doubt from seized Walther equipment. You are likely to encounter this pistol or a copy of it in your travels, and an appreciation of its strong and weak points is recommended.

Obviously, a pistol is only a cartridge launcher, so it is only as good as the cartridge. In my judgment, .32 ACP is *not* a real stopper, and although a .380 can be loaded "hot," you can never make a .380 equal a similarly loaded 9x19mm.

The sights are hard to see for target work because of the small size and narrow notch. The front sight on the cinema range was small, low, and indistinct, while the rear sight was small and difficult to pick up quickly and index in the dark.

The single-action trigger was good on the example tested, but the double-action pull was quite heavy and tended to stick before suddenly giving way without warning. My slim, size-9 hands tend to get cut by the slide going back and forth. It hurts, and I imagine it hurts people with fatter hands more. Everyone I know who shoots a Walther PP/PPK has sliced hands. The nonselectable double-action trigger style becomes a liability after the first shot is fired, and you need to move. You are faced with dropping the hammer and starting again with a heavy double-action pull for the next shot, or running with a loaded handgun with the safety off in hand or manually recocking. The CZ 83 solved this problem in a similar-size package.

Even though the grip is short, and I can only get two fingers on it, it feels good in my hand.

The size and weight of the pistol are nice, but the .32 ACP caliber is unacceptable. You'd be better off adopting the Hungarian RK 59 for the better caliber. Although the test model was a steel-framed pistol, the Walther PPK is available in an alloy frame in .22 and .32 calibers. I'd prefer the lighter alloy gun if I were to adopt the .32 ACP weapon. Even with the steel frame and using factory 71-grain .32 ACP ammunition, the Walther PPK seemed to kick as much as a 9x19mm Glock 17.

The biggest drawback to this weapon is the caliber. Perhaps if THV ammunition were available in .32 ACP, all these lightweight, fairly small .32 ACP pocket loaders would be acceptable, but until that happens they are all limited by their power when considering them for combat situations.

Author testing the Walther PPK 7.65mm.

The 50-foot test target with the Walther PPK.

Walther PP/PPK

Right side of Walther
PPK 7.65mm.

Left side of Walther
PPK 7.65mm.

Walther P.38
9x19mm
(9mm Para)

SPECIFICATIONS

Name: Walther P.38

Caliber: 9mm Parabellum

Weight: 2.1 lbs.

Length: 8.6 in.

Feed: 8-rd., in-line, detachable box mag.

Operation: Blowback, auto.

Sights: Front blade; rear blade

Muzzle velocity: 950 fps

Manufacturer: Walther Waffenfabrik

Status: Army and police use

As the German army began growing in the 1930s, it started looking around for a new service handgun. It had begun World War I with the Luger P.08 as its new service weapon, but soon depleted its arsenals of this handgun. By war's end, the German military had used all sorts of substitute weapons, from M1879 10.6mm revolvers to foreign .32 ACP self-loaders. Obviously, this was not an ideal arrangement. Further, the German engineers knew that the P.08 had proven unsatisfactory in the mud of France, was extremely time consuming to make in terms of machining and skilled labor, and was not safe.

The Walther Company had begun promoting its double-action self-loader in 1929. Although not an original design, the Walther factory certainly made a lot of them. However, the German military did not want a .32 ACP or .380 ACP weapon: it wanted a 9x19mm. Walther made vari-ous test models, and finally in 1938, the German military adopted the Walther HP. It evolved into the P.38.

It was the first issue handgun in 9x19mm to be available with a double-action trigger. This represented a breakthrough. During the war, people who had never seen a Czech "Little Tom" double-action pistol or never realized that a double-action .32 ACP autoloader was available came in contact with the P.38. Instinctively, people liked the idea of a hammer at rest and no safety to forget. During the war, a P.38 became a high-quality trophy among the weapon-wise, while the less sophisticated preferred the Luger.

Although less complicated to make, the P.38 never completely replaced the Luger, which remained in production until 1943. And as happened during World War I, it turned out that the military needed a lot more handguns than anticipated. Once again, the various .32 ACP autoload-

187

ers of assorted vintage and lineage appeared. Still, for World War II, the P.38 was the first-line combat handgun of the German army.

Sights on the P.38 leave much to be desired. The front is a European barleycorn type, and the rear has a shallow U-shape blade. Holding both elevation and windage is difficult with this type sighting. On the cinema range, the front sight is quickly picked up (though it would be better if it were white), but triangulation is difficult because of the rear-sight setup.

The pistol is wide for its cartridge because the recoil springs are situated on the outside of the barrel rather than under it. Although this might not make any difference for a military weapon, it does make the weapon bulkier than it needs to be.

The safety can be applied to load the weapon, so the weapon does not have to be handled "hot." Some decockers on wartime guns have been known to be faulty, so watch out for this and always point the weapon in a safe direction. When I was a kid, no one carried a double-action pistol like the Walther P.38 with the safety on. The safety was viewed only as a decocker, and the lack of a safety (and the resultant ability to simply pull the trigger and fire the first shot) was viewed as one of the good points of this design. Now, many knowledgeable people carry their double-action pistols that have slide-mounted decockers/safeties on safe. This is especially common with police officers because of the weapon-snatch problem. I wonder if the Germans who carried P.38 (or PP/PPK pistols) in occupied territories carried them on safe. I would imagine they carried them off safe. I have never seen a holster where the leather had an impression on it showing an on-safe carry, but I have seen a lot showing that it was carried off safe.

Although the P.38 does not have a magazine safety, it does have a loaded-chamber indicator. The magazine release is located in the butt, which necessitates using two hands for release and makes it slower.

Accuracy varies depending on how well the barrel and locking block are fitted to the frame. The single-action pull on the tested example is quite short and light, while the double-action pull is heavy with a lot of slack in it.

On the positive side, the weapon is easily stripped for field cleaning. The grip is comfortable, and the weapon was designed to be used by left-handed as well as right-handed people. The weapon is widely available throughout the world; you are just as likely to find one in Asia as in Africa. It is quite a good weapon and one that can be used with confidence. You would do well to acquaint yourself with its features should your paths cross at some point.

The Walther P.38 being carried by a member of the French Forces of the Interior during World War II.

German airborne troops armed with Luger or P.38 pistols to back up the MP40 (left).

The P.38 in action (below).

The P.38 with 50-foot test target.

Heckler & Koch
VP70

9x19mm
(9mm Para)

To my knowledge, this weapon has never been adopted by any country as a military weapon. It has, however, found a home in various special operations units throughout the Western world, in some African military units, and in a variety of quasi-police agencies.

It is typical of a number of weapons that do not appear to be very good on the formal target range or in the gunshop, but that shine in combat situations. It is far superior to the more conventional Beretta M92 (M9, M10), but not quite as good as a Glock 17. It is certainly among the top five military handguns.

On the formal range, the trigger pull was quite difficult, and it made it hard to get a tight formal group. The pull felt like that of a staple gun, in fact, and not at all like a conventional double-action trigger—and certainly nowhere as good as the Glock 17 is. This problem totally disappeared

SPECIFICATIONS	
Name:	H&K VP70
Caliber:	9mm Parabellum
Weight:	1.8 lbs.
Length:	8.03 in.
Feed:	18-round box magazine
Operation:	Blowback, self-loading DA
Sights:	Front blade; rear notch
Muzzle velocity:	1,170 fps
Manufacturer:	H&K GmbH
Status:	Obsolete

on the cinema range, where the weapon shot quickly and accurately. The smooth outline of the weapon made it easy to carry and produce. Accuracy, if the shooter does his part, is good, no doubt because of the fixed-barrel system.

On the night I tested this weapon on the cinema range, I also tested a new Smith & Wesson M5906 pistol. The VP70 was substantially better as a combat pistol than the M5906. Frankly, it is amazing to me that the Europeans are able to build semiauto pistols that are so much better than the average U. S.-built semiauto.

The sights on the VP70 are unusual in that they use shadows to get the desired effect. I would have preferred a simple white set of sights, but these are quite good. Even though it is a double-column-magazine pistol, the plastic grip feels good in my hand because the plastic allows for a slender grip. In this regard, it is much like the Glock 17.

I really like the VP 70. Previously, I had only used this weapon in the full-auto mode with a stock since it is designed to fire burst when the butt stock is affixed, and thus I never had an opportunity to view it as a pistol. It is a well-designed combat pistol, and, although too big for the average infantryman to use as a backup to his rifle, it is an excellent choice for soldiers who do not carry shoulder-fired weapons. It is an accurate, safe, high-capacity handgun. If the weapon is equipped with a stock, and the soldier is taught to keep the selector in semiauto mode, it should give him the ability to engage man-targets at 200 yards because of the 9x19mm cartridge's flat trajectory. Then if things really get bad, you always have the three-round-burst feature to use against close-range targets that need to be saturated with bullets. The VP70 is a very underrated weapon that puts people off unless they take the trouble to use it as it is meant to be used.

Right view of Heckler & Koch VP70 9x19mm.

Left view of Hechler & Koch VP70 9x19mm.

Front sight on the Heckler & Koch VP70. Note the unique construction of the front sight.

The 50-foot-test target shot with the Heckler & Koch VP70.

Walther P5
9x19mm
(9mm Para)

This pistol is an outgrowth of the wartime P.38 pistol. In the 1970s, German police departments decided to modernize their weapons. Specifically, they wanted a 9mm handgun suitable for uniformed and plainclothes officers in response to the increasing terrorist problems. Some of the specifications of the pistol were that it have no safety, hold at least eight rounds of ammunition, fire 9x19mm cartridges, and be within certain weight and length requirements. Three weapons met the criteria: the P5, the SIG P225 (P6), and the Heckler & Koch PSP (P7). The P5 is perhaps the most traditional of the weapons and has been adopted by some of the local police forces, but it is well behind the P6 in popularity. Perhaps cost is one reason: the P5 is about twice as expensive as the P6. Despite this high cost, the Dutch police recently bought 35,000 P5s, and these pistols are found in various elite paramili-

SPECIFICATIONS

Name: Walther P5

Caliber: 9mm Parabellum

Weight: 1.8 lbs.

Length: 7 in.

Feed: 8-round, detachable box mag.

Operation: Falling block (modified P.38)

Sights: Front blade; rear square notch)

Muzzle velocity: 1,146 fps

Manufacturer: Walther Waffenfabrik

Status: Current production

tary police units worldwide.

The slide on the P5 is solid on top, unlike the P.38 slide, thereby avoiding the cracked-slide problem common with the latter. The slide decocker has also been replaced with a decocker on the receiver that is quite handy to use—in fact, much more so than the P6 decocker. The sights are quite easy to pick up on both the formal and cinema ranges, though I think the rear would be better with dots on each side rather than the middle-placed stripe.

My test on the formal target range produced a 1 7/8-inch group size, with four of the five in a 1 1/4 inch group. This accuracy was due in part to the excellent trigger. The P5 has a built-in trigger stop with no over-travel and has a good surprise break.

It is easy to load and drop the hammer. The double-action pull is quite good and does not seem to be as long in reach as many double-action pistols. The heel-mounted magazine release slows reloading down, but reloading is not generally that critical an issue on a military pistol. The magazine is single-column, thus limiting capacity, but for real-world purposes, it is probably sufficient and does help avoid the "big-butt" problem common on conventionally framed self-loaders.

The P5 has an aluminum receiver and is lightweight, although heavier than the Glock. Even so, recovery is quite fast, and repeat shots can be made with ease. Although the slide is quite broad (a disadvantage for concealed carry inside the waistband), this is irrelevant for a military pistol. Still, it does not present the thin, elegant silhouette of the French M1935A or SIG P210. Instead it looks much more Germanic, even Rubenesque.

Overall, the P5 is large for its caliber and neither as safe as the Glock nor as light. These weapons are rare in the United States (I tested the only one I knew of), but they are more common in Europe. If you happen upon one, you could do a lot worse than it.

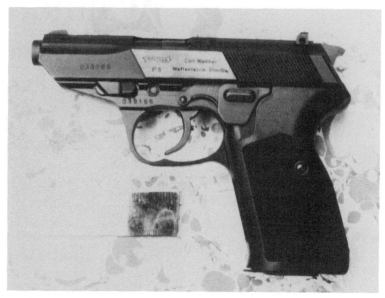

Left view of Walther P5 9x 19mm.

The receiver-mounted decocking lever that allows the shooter to decock this P5 without breaking the grip is an improvement over having the slide-located safety found on the original P.38 design.

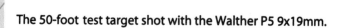

The 50-foot test target shot with the Walther P5 9x19mm.

SIG-Sauer P225

9x19mm
(9mm Para)

This pistol, also known as the P6 in Germany, was designed in response to the German police trials of the 1970s to find a new 9mm handgun for uniformed and plain-clothes officers (discussed in the section on the P5). Basically, the new weapon had to meet four criteria: have no safety, hold at least eight rounds, fire the 9x19mm cartridge, and be within certain weight and length requirements.

The P225 is an excellent weapon. Because of the single-column magazine, the grip on the P225 is sufficiently shallow and narrow to allow you to get a good grip on a weapon (unlike the P226), and it has a good feel to it. I had no problems whatsoever using the hammer drop on this particular weapon with my thumb. The magazine tested with this weapon was not functioning properly and caused some premature slide stops, but that had nothing to do with the handgun itself. The P225's muzzle flash

SPECIFICATIONS

Name: P225 (P6)

Caliber: 9x19mm

Weight: 1.6 lbs.

Length: 7.02 in.

Feed: In-line single column

Operation: Recoil

Sights: NA

Muzzle velocity: NA

Manufacturer: Sauer

Status: Current production

in the ammunition I was testing (Fiocchi 9mm) was the same as with the SIG P226, despite the P225's shorter barrel.

The front sights on the P225 are black and cut at an angle so that on the cinema range the sight appears gray. This causes the sight to disappear and prevents good indexing; it can be corrected by painting the sights white.

The P225 is a nonselectable, double-action automatic, and this causes some problems for some users. For military purposes, however, it is probably more of a solution than a problem. It does not have a magazine disconnector, which is unfortunate for a military handgun. If the sights were changed to duplicate the sight picture of the P220/226 and a magazine disconnector were added, this would be an ideal military weapon. Night sights are available for the P225; on military weapons, I suppose, they're standard.

All in all, I would rate this pistol very highly.

It is lightweight, safe, quick, and highly accurate. I got five shot groups of 3 5/8 inches; one of those was a flyer, and the rest would cover between 2 and 2 1/2 inches. For military purposes, I would rate this as number two after the Glock 17 in 9mm pistols that I tested.

German 9x19mm police pistols: (top) SIG P6; (bottom) Heckler & Koch P7; (left) Walther P5.

The SIG-Sauer P225 (P6) 9x15mm being tested by the author.

The 50-foot test target shot by the SIG-Sauer P225 (P6) (above).

A top comparison of the three standard German police handguns in 9x19mm: (left) Walther P5; (middle) SIG Sauer P6; (right) Heckler & Koch P7.

Heckler & Koch P7

9x19mm

(9mm Para)

The P7 was also designed in response to the German police pistol trials of the 1970s. Among the designs submitted, the P7 was perhaps unique because it has its cocking mechanism located in the weapon's grip.

This pistol has no conventional safeties of any type, per the police requirements. Grasping the weapon firmly cocks the weapon; letting go of the front strap allows the weapon to be decocked and thus completely safe. When you first use this pistol, it appears to be, on a formal range anyway, a quite safe design. However, upon close examination, this pistol loses some of its appeal. In my opinion, this is a dangerous handgun and should be avoided.

The trigger is generally consistent, offering only a single-action firing technique, and thus there is no shifting of the trigger system as you would have with a conventional double-action pistol.

The size of the pistol is also quite pleasant for

SPECIFICATIONS

Name: H&K P7

Caliber: 9x19mm Parabellum

Weight: 2.7 lbs.

Length: 6.54 in.

Feed: 8-round, detachable box mag.

Operation: Blowback w/gas-piston retard.

Sights: Front blade; rear notch

Muzzle velocity: 1,365 fps

Manufacturer: Heckler & Koch

Status: Current production

---

a 9mm handgun, although, because it is made out of steel, it is really too heavy.

Because of the fixed barrel, accuracy is quite satisfactory on this weapon. I placed five shots within 2 inches overall, with three of those shots going into 3/4 inch.

The sights were quite good both on the cinema range and the formal target range, but particularly so on the cinema range. The large, white front and rear sights facilitate good indexing and a quick response in a dark range situation.

The P7 can be used in either hand quite readily, a desirable trait given the fact that 11 percent of the population is left-handed. It fires from a locked position, but it still seems to kick as much as or more than a .45 automatic.

I have had four of these pistols break on me, and I have some real questions in my mind as to whether or not these pistols are designed for long-term use. The police trials required the pistols to meet a 10,000-round test, and that is probably more than most people put through them. However, my experience has been that the P7 tends to break the cocking mechanism or the firing-pin mechanism after about 3,000 rounds of full-bore European 9mm ammunition. I understand from some secondhand reports that now instead of fracturing, the mechanisms tend to be battered out of dimension, causing full-auto fire. The example I tested had fewer than 7,000 rounds fired through it.

The P7 has some considerable drawbacks. The worst thing about it is its unsafe design. As you pull the pistol out of the holster, you automatically cock it, and you are left with a cocked pistol with no safety and a very light trigger pull. If this were the only weapon being used, as might be the case in some military or police situations, careful training might make it a little less unsafe. Finally, it is also too heavy for its size.

I cannot recommend this pistol. Much better, lighter, and safer designs are available that are just as accurate and powerful, such as the Glock 17.

Right view of Heckler & Koch P7 9x19mm.

Left view of Heckler & Koch P7 9x19mm (above).

The 50-foot test target shot with Heckler & Koch P7 (left).

Colt Army Special

9x29mmR

(.38 Special)

The Colt Army Special is included in this survey because it was a standard handgun in the Greek army and navy in World War I, and because 10,000 of them were purchased by the French government during that war. Its successor, the Official Police (also known as the Colt Commando), was used in its militarized version during World War II.

Despite the exotic name, this Colt handgun saw very little use during World War II; it was primarily used by plant security guards, police officers, and the U.S. Federal Bureau of Investigation, especially in the 2-inch-barrel version.

The Army Special is a .38 caliber handgun with a steel frame. Although not quite as durable as those on the New Service, the Army Special's parts are well made of good material, and they are subject to low stress because of the relatively low power and recoil of the .38 Special cartridge.

SPECIFICATIONS

Name: Colt Army Special

Caliber: 9x29mm (.38 Special)

Weight: 2 lbs.

Length: 11 1/4 inches with a 6 inch barrel

Feed: Revolver

Operation: Double action

Sights: Blade/notch

Muzzle velocity: Standard for cartridge

Manufacturer: Colt

Status: Obsolete

The Colt Army Special suffers from many of the same deficiencies as the Colt New Service.

First, its grip is large (though not quite as large as that of the New Service), and it does not fit the hand well without an adapter. When rapid double-action strings are fired, it tends to shift in the hand somewhat.

Second, although the trigger pull on single-action is quite good, on double-action it is heavy and tends to stack. Interestingly enough, the same trigger action is found on the Python, which can be modified to be the best double-action pull obtainable in a revolver. It would indeed be interesting to see if some old Greek Army Special revolver with a battered exterior could be modified to equal a Colt Python trigger.

Third, all Colts offer a cylinder-release mechanism that I find awkward. You must pull the release to the rear rather than simply pushing it forward, as you do on a Smith & Wesson. Years of shooting Smith & Wesson revolvers have convinced me that merely pushing the release in and flipping the cylinder out is a faster procedure than pulling the release back and pushing the cylinder out.

Fourth, I dislike the fact that the extractor rod does not lock up. I believe this is a weaker system that is subject to damage if struck hard.

Fifth, the front sight is dark, although it does line up well in the dark because it is high on the barrel, thereby aiding indexing. The rear sight is low and does not offer much aid for quick triangulation in the dark.

The Greek Army Special revolvers all appear to have led a hard life. The very few that are encountered all seem very battered. Of course, life in the Balkans was always hard on weapons, and the Colt Army Special is no exception. The condition of the remaining examples certainly establishes that the weapons saw much use. The Greek or French military man who was issued a Colt Army Special received a satisfactory weapon, within the limits of the caliber. The design lacked some useful features, such as the swing-open frame of the Rast & Gasser. I would prefer a Webley MK VI myself, due to the better caliber and grip, but certainly the Army Special was better than many military revolvers found in holsters during World War I.

The author firing the Colt Official Police 5-inch revolver.

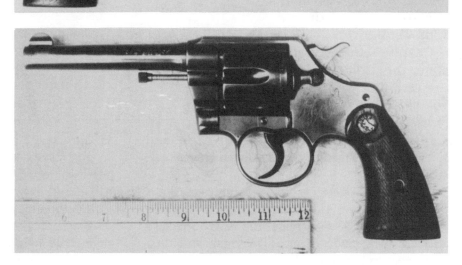

Right side of Colt Official Police .38 Special.

Left side of Colt Official Police .38 Special.

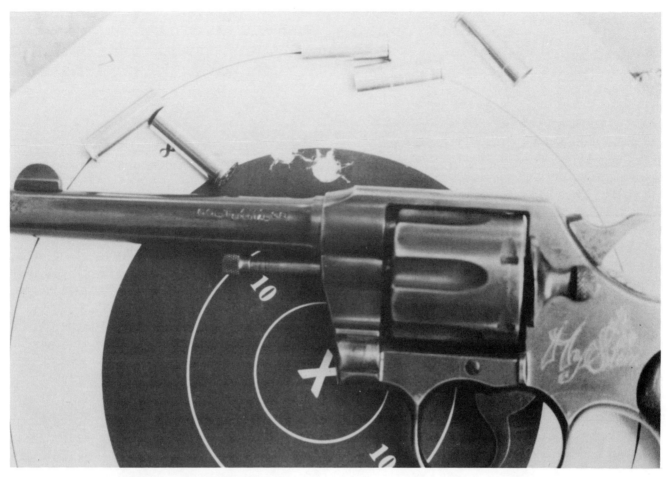

The 50-foot test target with the Colt Official Police.

M37

7.65x17mmSR
(7.65mm Browning, .32 ACP)

There is nothing spectacular about the Hungarian M37. The example I tested was a standard model without the later German army side-safety modification, so it had only a grip safety. I can understand why the Germans modified the safety.

The hammer is small and shrouded by the slide and length of the receiver. As a consequence, the M37 is difficult to cock rapidly, and you are forced to either trust the grip safety or carry the weapon chamber empty and load the weapon immediately prior to firing. Otherwise, since you must cock the weapon, you have to break your firing grip, thereby covering the rear of the sight and further increasing your problems. Of course, you could cock the pistol with your weak hand, but then you would not have both hands to work the slide.

Although the grip safety did seem to work well, I would not carry the weapon cocked and

SPECIFICATIONS

Name: M37

Caliber: 7.65mm (.32 ACP)

Weight: 1.6 lbs.

Length: 6.8 in.

Feed: 7-rd., in-line, detachable box mag.

Operation: Blowback, semiauto

Sights: Front blade; rear V notch

Muzzle velocity: 984 fps

Manufacturer: NA

Status: Obsolete

loaded. Even if you delay loading it until you're ready to fire, for subsequent shots you have a cocked pistol in hand. Moving with such a weapon is dangerous, and you must concentrate on holding the weapon *loosely* in your hand because a firm grip disengages the grip safety. In tense shooting incidents, an accidental discharge could easily occur if your finger rested on the trigger. I suppose, if you are properly trained to hold the pistol loosely, with your finger off the trigger, it would be safe to carry.

In fairness, a firm grip is needed to deactivate the safety; in fact, it does require quite a bit more pressure to disengage than the grip safety on the Colt Government Model. Still, a safety system that is designed to respond to the instinctive actions in a shooting situation—and not be dependent on an individual remembering how to grip the weapon—would be better. This is especially apparent when you consider the variety of people who use the weapon—some properly trained, some not.

The heel-butt-mounted magazine prevents rapid reloading, and the single-action trigger system adds to the weapon's unsafe carry and sluggishness problems discussed above.

The front sight is dark and small, and the rear is small, so proper indexing is difficult. A dab of white paint would aid indexing a lot. The grip is very straight, and the weapon tends to point low, causing low bullet strikes. I found this interesting because I practice extensively with the Colt Government Model, which is also straight, but apparently not as much so.

The only positive things that can be said are that the M37 functioned perfectly during the firing test and recoil was light, as might be expected in a steel-framed, single-column, single-action, .32 ACP self-loader.

Last, the M37 suffers from caliber limitation. It is just as heavy as a Glock 17 but in a much smaller caliber. Even in .380 it is not powerful enough with conventional loads. Many better weapons are available in either of the two calibers.

Right side of Hungarian M37 7.65mm pistol.

Left side of Hungarian M37 7.65mm pistol.

Close-up of the grip area of the M37 illustrates the difficulty of cocking the weapon.

The 50-foot test target for the M37.

RK-59

9x18mm
(9mm Makarov)

The RK-59 is the lightest, most powerful PPK ever made. This excellent military handgun is issued to members of the Hungarian military and police forces.

Although the RK-59 is the size of the Walther PPK, the recoil shoulder is designed to be slightly higher over the web of the hand, and thus all the hammer bite and slide slashing so common with Walthers are absent. Even better, this pistol has an alloy frame. In the Walther PPK series, only .22, .25, and .32 pistols have alloy frames; in .380 you get steel. Fortunately the Hungarians do not think an alloy frame on the .380 a problem. Thus, with the RK 59, you get a very lightweight pistol that will not weigh your pocket down. More good news: the pistol is chambered for 9mm Makarov, which is better than .380 by at least 10 percent.

The bare-metal finish on the alloy frame looks nice for civilian use, and the blued slide is also

SPECIFICATIONS

Name: RK-59

Caliber: 9mm Makarov

Weight: 1.3 lbs

Length: 7 in.

Feed: Single-column magazine

Operation: Blowback

Sights: Blade/notch

Muzzle velocity: NA

Manufacturer: State factory

Status: Current issue in Hungarian police agencies and military units

acceptable for similar uses. If adopted by a Western military, the gun would likely receive an anodized dark matte finish. If it were not for the fact that this pistol is a prized collector's piece, I would consider applying Robor's NP3 to it to improve rust resistance and make it tactically more acceptable.

Ammunition in 9mm Makarov is always a bit of a problem. For unarmored subjects, you want expanding ammunition, but for military purposes, a THV-type round would be ideal. Perhaps some type of steel-core penetrator round, as is common with the 7.62mm Tokarev or the PSM, would be useful. Generally, the Soviets were always favored such steel-core rounds. In the West, we always assumed that was so they could make them cheap and fast, but I am not so certain. I think it may have been to better counter Western armor technology. In any event, such ammunition, if available, would certainly be useful in this weapon.

On the formal range, the double-action pull was typical of a Walther PPK: very heavy with a lot of stacking. No one will ever confuse a Walther PPK with a Smith & Wesson pre-war Magnum if tested blindfolded! The single action was quite crisp but suffered from the typical autoloader problem of substantial overtravel. The sights are small, dark, and hard to pick up, and, as a consequence, I was able to achieve only a 3 3/4-inch group at 50 feet. A dab of paint would go a long way to improve the sight's visibility, which should increase accuracy accordingly.

On the cinema range, the weapon performed much better. It was quick to put into operation and fast to shoot. Indexing was slowed a bit by the low, small, dark front sight, but a little paint on the front sight would allow the eye to pick it up more rapidly and distinguish it from the darkened slide, thus facilitating indexing.

This pistol is an ideal size for the infantry trooper who is already overburdened with equipment. Although it does suffer from the typical double-action slide safety mounted problem (noticeably absent on the CZ 83), the combination of size, weight, and caliber adds up to a highly rated weapon for field troops. If it does not make the top-five list of available military handguns, it certainly gets into my top 10.

The author testing the RK-59. Note the muzzle flip that occurs with 9mm Makarov ammunition because of the gun's lightweight alloy frame.

Right side of RK-59 9x18mm pistol.

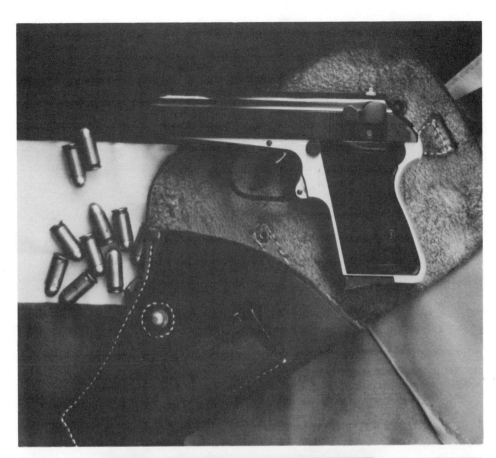

Hungarian RK-59 with holster.

The 50-foot test target for RK-59 9x18mm pistol.

Ruger P85
9x19mm
(9mm Para)

The Israeli air force recently adopted this weapon as its new service handgun, and it was also a serious contender to replace the Beretta M92 in the U. S. military. Fortunately for the U.S. forces, it was not adopted. As for the Israeli forces, well, it is my opinion that they really do not understand handguns. They carry their P85 pistols in condition three even in their elite units, a condition no elite unit in the United States or Britain would tolerate. I really do not think the Israelis consider handguns serious fighting tools. They do not have a history of using the combat handgun to draw on, so their selection cannot be given much merit.

This P85 is basically a conventional double-action, all-steel, double-column-magazine weapon with no particularly outstanding design features. The pistol that I tested had never been carried and had been fired fewer than 100 times,

SPECIFICATIONS	
Name:	Ruger P85
Caliber:	9mm Parabellum
Weight:	2 lbs.
Length:	9.4 in.
Feed:	Double column
Operation:	Recoil
Sights:	Front blade; rear notch
Muzzle velocity:	NA
Manufacturer:	Ruger Firearms
Status:	Standard air force weapon in Israel

but the finish was already chipping off the right side of the frame. I am not a "nut" on durable finishes, but I cannot help but think this reveals a serious weakness somewhere in the finishing of the weapon.

It is built like a locomotive, but this feature also creates excessive bulk. All the edges of the weapon are sharp, and they tend to cut the hand when the weapon is handled rapidly. The grip is somewhat cramped because you must force your hand up to the trigger guard to get your hand high enough to minimize muzzle whip and shift. In so doing, the middle finger is pressed so hard around the trigger guard that it gets rapped when the weapon is fired, causing some discomfort. This problem needs to be fixed, as Smith & Wesson did on its third-generation autoloaders.

On the cinema range, the two rear-sight dots were too small for rapid pickup, and the front sight was much too small to pick it up and put it on target. It also tended to blend in with the background, although it is better than all-black sights.

The double-action pull is quite good, and the transition from self-cocking mode to single-action mode is pretty smooth. The single-action pull is quite light and would be dangerous in undertrained hands; I fear that unintentional discharges will be common with this weapon.

It is difficult

Right side of Ruger P85 9x19mm.

to apply the safety after the first shot because of the grip's bulk. It requires that you shift the weapon so the thumb can depress it, unless the weak hand is used, which obviously is unacceptable. Also the safety system is like all those slide-mounted decockers in that it requires the shooter to apply it and disengage it, leaving him with a heavy double-action pull for the next shot or an off-safety cocked pistol with a light, single-action pull. Most people, unless very well trained, will opt to leave it cocked to make shooting quicker, and these same low-skill people will likely move with their fingers on the light trigger. A better, though not totally safe, system is found on the CZ 75-type pistol. The CZ 75 system requires that the weapon be "hot" when loaded, and then the hammer is manually dropped on a loaded round. The Taurus method is too complicated, in my opinion, so this may be one of those take-your-choice-of-bad-alternatives situations.

The magazine release is small, and the square shape is difficult to disengage. The lanyard loop on the butt is a nice feature, but, unfortunately, it is so large that it prevents you from speed-reloading unless you put a pad on the butt of your magazine. Pushing the magazine up with the heel of your hand will cause your hand to strike the lanyard loop, not only hurting your hand but also failing to firmly seat the magazine.

One of the P85's worst features, however, is the single-action trigger. You must release it to go forward after a shot. There is a tendency to short-stroke the weapon, and this will prevent fast repeat shots. With practice, one could overcome this problem, but since it has never happened with any other pistol I've fired, it's worth noting this design flaw.

The Ruger P85 is a tough weapon constructed in old-fashioned steel, but it is no tougher than the M92 (M9, M10) Beretta or the SIG P226, and certainly not the equal of the Glock series.

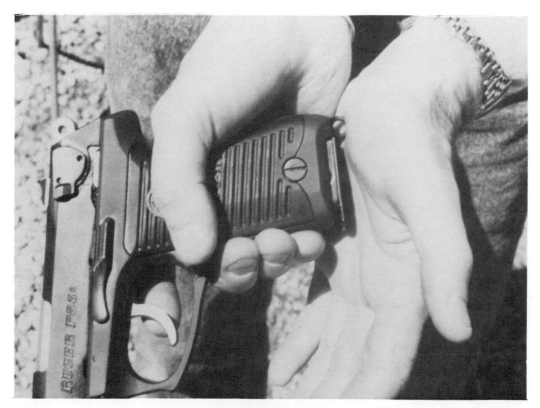

One problem with the Ruger P85 is the placement of the lanyard loop, which prevents the shooter from firmly seating the magazine with the heel of his hand.

The 50-foot test target fired by the author with the Ruger P85.

Model 1889 Ordnance Revolver

10.4x20.2mmR
(10.4 Italian Revolver)

I remember back in the early 1960s when these revolvers were available for $9.95 from Ye Olde Hunter in Alexandria, Virginia. Despite the price, not many people bought them, and of those that were purchased, many suffered the final indignity of being made into lamps.

Although I suppose that at one time they were plentiful, I had a hard time finding an example to test. Fortunately, I was able to buy one, and Fiocchi imports the ammunition. That tells me that these revolvers must be plentiful somewhere. Italy? Greece? The former Italian colonies in North Africa? I don't know, but I am glad they make the ammunition because it saved me from having to do so. It also confirmed the importance of including this weapon in this test; you may come across it in your travels. (As an aside, Fiocchi ammunition has a lot of flash to it, so remember this characteristic before you discover it in a dark alley

---

SPECIFICATIONS

Name: Model 1889 Ordnance Revolver

Caliber: 10.4x20.2mm

Weight: 2.2 lbs.

Length: 10.25 in.

Feed: 6-rd. revolving cylinder

Operation: DA

Sights: Front blade; rear V notch

Muzzle velocity: 840 fps

Manufacturer: Glisenti

Status: Obsolete

---

some night and think your weapon blew up. I thought my M66 Smith & Wesson had blown up the first time I shot Federal .357 Magnum 125-grain JHP loads in it in the dark.)

The Italian Ordnance Revolver has light recoil, but double-action shooting is slow because of the heavy pull. Additionally, the lever on the left rear side of the revolver that facilitates takedown has a tendency to strike the thumb, dig in, and cause some pain. The rear sight is narrow and shallow, while the front sight is too large for rapid indexing. The Model 1889 I tested was the variant with the folding trigger and no trigger guard, a configuration that caused some problems using a two-hand position. Of course, this would not be a problem with one-handed shooting. On the plus side, this design does allow you to stick the weapon in your pocket and withdraw it rapidly without fear of striking any projections. I have seen Webley pocket revolvers with that same design, and the Patterson by Colt incorporated a similar design. Although at first it seems odd and ill-conceived, it really is not so bad.

The trigger features a smooth face, as is common today with "combat-style" triggers. The smooth surface allows the trigger to move even while pulling a heavy double-action trigger, which is better than the target shooter's grooved trigger. Despite the ropelike grip of the weapon and loads that are as powerful as .455 Webleys, the weapon did not shift in my hand, even when firing rapid double-action strings.

Reloading is slow because of the ejector rod that had to push each empty out. Unlike on the Colt SAA design, the rod is not attached to the barrel nor is it spring loaded. Thus, you have to swing the rod into place, pull the empty out, pull the rod to the rear, index the cylinder, and repeat. The cut in the frame that fits the shape of the empty had to match perfectly or it would not eject. This would make reloading especially difficult at night. The cut should be slightly wider to allow for slight misalignments and avoid jamming the weapon.

Made in 1925, the weapon tested was a very poor example of the combat revolver, especially when compared to the French M1873 11mm or the Webley revolvers that were its contemporaries both in vintage and caliber. Still, it is a much better fighting weapon than the 9mm Glisenti "Brixia" that replaced it.

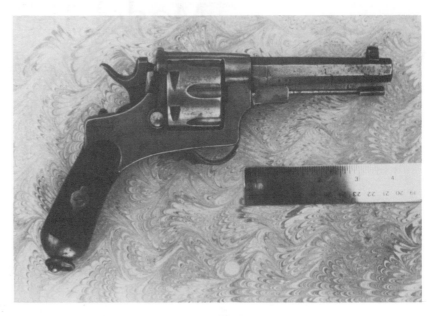

Right side of Italian 10.4mm Ordnance Revolver.

Left side of Italian 10.4mm Ordnance Revolver with sideplate removed. Turning the lever allows the sideplate to be removed quickly, thus facilitating keeping the weapon clean.

The author unloading the
Ordnance Revolver.

The issue holster for the
Italian 10.4mm Ordnance
Revolver.

The 50-foot test target for
the Ordnance Revolver.

M1910 Glisenti "Brixia"

9x19mm Glisenti
(downloaded 9mm Para)

The Italian military has had some real difficulties in selecting handguns. This is really surprising when you consider the tradition of fine Italian gunmakers, such as Beretta, and the number of gunmakers there today who make fine copies of Colt SA Army pistols and Czech CZ 75s. But prior to World War I the military had an ongoing problem. Its ordnance revolver was certainly no better than other revolvers available during the period, and worse than a number of them. The Beretta M1915 has numerous problems, and even the Beretta M1934, which is certainly an adequate pocket pistol, but does not meet the requirements for a military pistol, in my opinion.

The problem did not cease when the Italians adopted the Glisenti, or Brixia model. The pistols are similar except the Brixia lacks the grip safety found on the Glisenti and has a reinforced sideplate. Although the Brixia was a commercial pis-

SPECIFICATIONS

Name: Model 1910 Glisenti "Brixia"

Caliber: 9mm

Weight: 1.9 lbs.

Length: 8.1 in.

Feed: 7-rd., in-line, detachable box mag.

Operation: Delayed blowback, semiauto

Sights: Front barleycorn; rear V notch

Muzzle velocity: 1,050 fps

Manufacturer: Glisenti

Status: Obsolete

tol, some were bought for government trials and designated the "M1912." The cartridge is the 9x19mm Italian loading, which had slightly different dimensional specifications but was functionally interchangeable in most weapons. You must use care *not* to load standard 9x19mm ammunition into this weapon even though the cartridges are identical except for powder charge. This must have given ordnance officers nightmares, especially when they had other weapons in their inventory (such as the Beretta M38 SMG) that used the standard or even hotter 9x19mm loading. Although the ammo looks the same and will chamber in the other weapons, if you use the wrong one, you either blow up your pistol or have malfunctions in your submachine gun—which can get you killed. It would make an interesting story to uncover why Italian military handguns have been so bad. Perhaps the Italians simply did not have anyone around who knew the real uses of the military handgun, and thus there was no demand for a good weapon.

The Brixia pistol looks good, and when you take it apart you will note the engine-turned interior with approval. The grip has an angle to it, similar to the Luger Lahti, Heckler & Koch P7, and Glock, so it should do well on both the formal and cinema range. Sights are the typical European barleycorn type. Keeping your elevation on the formal range is demanding; however, on the cinema range, the sights do pick up fast since they get up in the air at the end of a slender barrel and allow good indexing. The rear sight is too shallow and narrow, making both formal and cinema range firing tedious. Both sights are dark, and pickup would benefit from being painted white.

The poor sights and trigger pull resulted in groups on the formal range of 4 5/8 inches, well over twice the average size. I do not know why the trigger pull was so gritty and hard since the parts appeared to be fitted together properly, and the whole surface was carefully polished. It must be a design flaw.

The most astonishing thing about the Brixia is its recoil. Now, I realize that most of you who have never shot one will laugh at this and think the author weak-handed. After all, shooting a .380-power-level cartridge in an all-steel pistol with a grip angle that traditionally is considered good should not any be heavy enough to draw comment. The recoil is so hard that shooting this weapon is a chore. Three other shooters were on the range the day I tested the Brixia, and after a couple of rounds, I was complaining about my fingers being numbed by the recoil, so they all tried it. Among them were Leroy Thompson, well-known author and executive protection/counterterrorist specialist, who fired once and stopped; Edward Seyffert, a local law enforcement agent and well-known police firearms trainer, who fired one shot; and another gentleman who used to put in poles (mostly by hand) for the Missouri Highway Department and broke horses in his spare time (his hands felt like a dull file and his arms were as big as my legs), who

fired twice and called it quits. Realizing that I was not exaggerating or developing a weakness, they all chuckled heartily at my predicament: they knew I had to shoot a proper five-shot test target and take it for a 20-50-shot tour of the cinema range. The level of comments went from mildly amusing to borderline obscene, and I will not bother you with them.

After a few more painfully fired shots, it dawned on me that I would never get a decent group fired that way, so I pulled my heavy wool glove out of my pocket, cut off the fingers, and shot my group and ran through my cinema test. It still gave my fingers a very strong knock, but at least they were not numb from the shooting.

I do not know why this pistol has such violent recoil. All I can conclude is that, when you fire, your trigger finger somehow gets knocked to the front of the guard and strikes it a mighty blow. Since that test (November 1988), at every gun show I've attended, I have asked all the dealers who had Brixias whether they had encountered the recoil problem. To date, I have been met by blank looks: not one person I have encountered has ever even shot one. In fact, when I inquired at the National Automatic Pistol Collectors annual meeting not only did I not get any reports of actual firings, but everyone looked at me as if they thought I was odd to shoot it.

I suppose the Italian officer who wore his thin leather gloves and used his M1910 Glisenti to execute natives by shooting them in the back of the head might not have been bothered by the problems I discussed here—but they will bother you.

If you are desperately looking for a handgun in the bazaars of Algiers and are offered the choice between a like-new Brixia along with factory-fresh ammunition or a beat-up, rusty Webley Mk VI with old but clean ammunition, pick the Webley. The Brixia easily makes the list of one of the five worst military handguns tested. Lest you think my comments are because of my using the wrong ammunition, they are not. The ammo tested was made to .380 specifications, not 9x19mm Luger power.

The glove is needed for firing the Brixia 9x19mm because of heavy recoil and the impact on the shooting finger. It would not seem that such a low-power weapon would need this precaution, but the Brixia was painful to shoot, and not just for the author. Other shooters tried the weapon and experienced the same discomfort.

Right side of Brixia 9x19mm.

Left side of Brixia 9x19mm.

The Brixia 9x19mm with 50-foot test target.

Beretta M1915

7.62x17mmSR/
9x19mm Glisenti
(7.65mm Browning, .32 ACP/
downloaded 9mm Para)

Although many of these weapons were produced during World War I, they are quite rare now. This was one of the earliest Beretta-designed service pistols, and it had a few design features in it that will later be found in the M92 Beretta. A close look at the slide and barrel design will establish that these weapons are at least cousins, if not more closely related.

This pistol was made in .32 ACP and 9x19mm. However, the Glisenti loadings had one-third less propellant than typical o8 German loadings. I imagine that a +P+ round would certainly stress the weapon beyond acceptable levels. Still, it is more powerful than the .380 and .32 ACP pistol commonly found during the period, but is not any larger.

The safety falls to your thumb fairly easily, although getting it back on is difficult. Unfortunately, the slide locks back on the last shot but closes when you withdraw the magazine. Not

SPECIFICATIONS

Name: Beretta M1915

Caliber: 9x19mm

Weight: 1.3 lbs.

Length: 6 in.

Feed: 7-rd., in-line, detachable box mag.

Operation: Blowback, semiauto

Sights: Front blade; rear V notch

Muzzle velocity: 960 fps

Manufacturer: Beretta

Status: Obsolete

only does this make it tough to withdraw the magazine, since you have additional pressure against it, but you now have to rack the slide to load the weapon. This obviously slows up speed loading. But if you pull the slide to the rear, lock it back, and then withdraw the magazine, not only do you increase your time in reloading, you also run the risk of losing your barrel: with the slide locked back, the barrel can be quickly removed by merely lifting it. This might be convenient for a quick, thorough cleaning, but it is also a good way to lose your barrel. On the range, I fired a group, locked slide back, dropped my arm to my side and walked forward to look at my group. When I raised my pistol, I was shocked to see the barrel missing. Fortunately, I was able to retrace my steps and find it. I doubt that I would have been so lucky were I running through a muddy trench on the Austrian border in 1916.

The sights are bad both front and rear, and the trigger is hard. In fact, the weapon is very similar in this regard to the French Ruby pistol but the fact that caliber is 9mm makes this a better selection in my judgment.

The author test-firing the Beretta M1915.

Right side of Beretta M1915.

The rear safety of the Beretta M1915.

This disassembled view of the Beretta M1915 shows the ease of barrel removal.

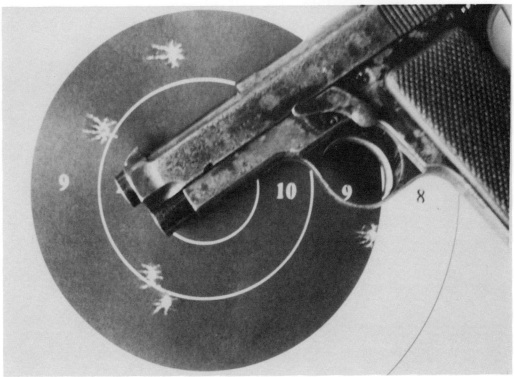

The 50-foot target fired by the author with the Beretta M1915.

Beretta M1919
6.35x15mmSR
(6.35mm Browning, .25 ACP)

This little pistol would seem out of place in a volume such as this one, except that it was at one time an issue handgun of the Italian air force. I assume that the weapon was merely a "badge of office" and not really designed to be a fighting handgun for downed airmen. However, perhaps the Italians felt that its light weight and small size made up for its caliber limitations. I recall reading that the German air force pilot who had killed more Russian tanks than any other man and who had been shot down a few times carried a Mauser .25 ACP pistol. It struck me as an odd choice since he appeared to have adequate space for something more effective. My choice would have been a Mauser M1934 7.63 with stock and a dozen spare 20-shot magazines; this would have given him better range, more power, more flexibility, and, if necessary, he could have used Russian SMG ammunition. However, he selected a .25 ACP.

SPECIFICATIONS

Name: Beretta M1919

Caliber: .25 ACP

Weight: 12.4 oz.

Length: 4.49 in.

Feed: Single-column magazine

Operation: Blowback

Sights: Blade/notch

Muzzle velocity: NA

Manufacturer: Beretta

Status: Obsolete

I have a sneaking feeling that no air force pilot is ever very good on personal weapons. Flying they may know, but I do not think they visualize themselves engaged in hand-to-hand combat. Apparently, it takes an infantry type to really see what handguns are for in the military. Rarely, unfortunately, are they asked.

As with all similar hammerless designs, the safety on the M1919 is difficult to disengage rapidly, worse to put back on, and dangerous to carry loaded with a round in the chamber without the safety on. Enough said.

Sights on this weapon are similar to those found on pocket pistols of similar caliber the world over. They are made small to either fit in your pocket (unnecessary when carried in a belt holster as the Italians did) or avoid tearing up your throat when your assailant takes your weapon away from you and jams it down your throat.

Unlike the Soviet PSM 5.45 mm pistol, no special ammunition is available to save this weapon from the apparent deficiencies caused by its caliber. About the best that can be said for the M1919 is that it would make an acceptable second gun or a primary one for an undercover man or woman. As a military weapon, it strikes out because it fails the critical test of stopping power.

Note the fired case exiting the Beretta M1919 as the author test-fires the weapon (above left).

Left view of the Beretta M1919 6.35mm (above right).

Right view of the Beretta M1919 and the 50-foot test target fired by the author (right).

Beretta M1934

7.62x17mmSR/9x17mm
(7.65mm Browning, .32 ACP/.380 ACP)

The Beretta Model 1934 is an all-steel weapon, single-column, butt-magazine-release pistol with a multitude of problems.

Even though it is all-steel, this weapon recoils quite heavily for its caliber. On the example that I tested, one round failed to fire for reasons that were not apparent to me. The sights are hard to see, the front is too shallow and the rear too narrow, the front is also pyramid in shape, which makes target work grueling.

My groups ran approximately 5 inches, roughly twice that of a Model 19. This was basically because the sights are so hard to use on the target range, not because of any intrinsic inaccuracy in the weapon. My first four shots went into approximately 3 inches, and the fifth took it out. I do not think that the method of holding the barrel to the side is conducive to accuracy, but certainly it is adequate for this type of weapon. However, it is difficult to shoot accurately.

SPECIFICATIONS

Name: M1938

Caliber: 7.65mm (9x17mm)

Weight: 1.3 lbs.

Length: 6 in.

Feed: In-line, single-column box

Operation: Blowback

Sights: NA

Muzzle velocity: NA

Manufacturer: Beretta

Status: Used by Italian police & military

The slide on the M1938 closes when the magazine is removed. Although the heel-butt-magazine release slows down reloading, once you disengage the magazine release and pull the magazine out, the slide will then close. Of course, this makes it a little more difficult to load properly. If you attempt to load this weapon with the safety on, it will jam. Therefore, you are forced to load it with the safety off, which is an undesirable trait for a military weapon. It would be nice to be able to load it with the safety in place when dealing with less well-trained troops.

The poor sights also are a drawback on the cinema range. They are too shallow, and they have no contrast, so indexing become arduous. On the formal target range, the recoil seemed to be quite heavy compared to the power factor; on the cinema range, the felt recoil really was negligible.

The safety was nearly impossible to disengage quickly with a strong-hand thumb. Because of its location, you have to push it up as far as possible with your right thumb, then shift your thumb and push off the remainder with the thumb—unless you happen to have thumbs that are normally issued to gorillas. This is very slow. If it is not all the way off, the hammer will sometimes, but not always, fire the weapon. For instance, if you pull your weapon out (cocked and locked), and push on the safety with your thumb but not flip it the whole 180 degrees, the hammer will sometimes fire the weapon. This is dangerous because

pulling the hammer back and clicking it again will sometimes cause it to fire a second time. The location of the safety is ineffective; it is too far forward, making it awkward for your thumb to hit and disengage it rapidly.

Although the M1934 is well made, it is not well designed for military purposes: it is slow to reload, difficult to use in low light, and not particularly powerful. Other available .380 weapons, such as the CZ 38/39 double-action series, are much better suited to the tasks of a military weapon.

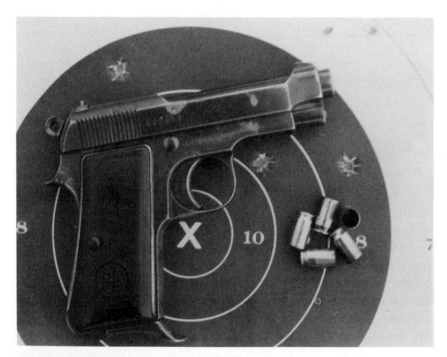

Right side of Beretta M1934 and the 50-foot test target fired with it.

Left side of Beretta M1934.

Beretta M1951

9x19mm
(9mm Para)

This is the current service weapon in Italy, where it has had little combat use. It is also a standard military handgun in Israel (M951) and Egypt (Helwan), where it probably has seen more fighting, but since neither of those armies is pistol oriented, its combat experience is still limited. Commercial Helwans made in Egypt by Maadi have been imported into the United States marked "Helwan."

Despite this lack of combat experience, this is actually quite a good weapon. The design is similar to that of the Beretta M92 and has shown itself to be quite reliable. Of course, the M1951 Beretta uses a single-column magazine, which features a single-action trigger style, but that configuration has some advantages. It has a straight-line feed, thus increasing its feeding reliability. The single-column magazine avoids grip bulk, and the single-action trigger makes it easy to shoot.

SPECIFICATIONS

Name: Beretta M1951

Caliber: 9mm Parabellum

Weight: 1.9 lbs.

Length: 8 in.

Feed: 8-rd., in-line, detachable box mag

Operation: Recoil, semiauto

Sights: Front blade; rear V notch

Muzzle velocity: 1,182 fps

Manufacturer: Beretta

Status: Used by Egyp., Ital., & Israeli armies

The magazine release is located on the butt, and that is always slow. It also requires two hands to use and has a tendency to get pushed off by car seats. The front sight is narrow and low, and the rear sight is small and shallow, thus indexing is slow. Painted white, they would show up much better. The sights and the gritty trigger on the Egyptian example tested combined to yield a 3 1/6-inch group on the formal range.

The safety on the M1951 is a crossbolt variety. This is unlike that found on most other combat handguns, and when you first see it, you will probably view it as awkward and slow. But you will be mistaken. The safety is one of the pistol's best features. It reminds me of the safety found on the Star Z-63 SMG, and I found that on both the Star and the M1951 you could flip the safety off and on rapidly without shifting your hand at all. In fact, it was faster to operate than a Colt Government Model. All you need to do is take up your normal firing position, with your right thumb (assuming a right-hand grip) resting with the knuckle on the button. Merely extending the thumb slightly will bump the safety off. To reengage, merely straighten the trigger finger out, flex the finger straight out, hitting the button with the inside of your knuckle, and it will flip on. In actual practice, I found it fast and easy. For left-handers, the procedure is reversed, but it is equally simple. You have no need for an ambidextrous safety, extended safety, or external safety, and since the pistol grips are flush with the safety, you avoid the problems associated with flipping the safety off while in the holster. I really like this safety system.

These pistols are not by any means common, as is a Colt Government Model or P-35. But because of the armies that have adopted them, their wastage in battle, and the poor accounting systems for small arms, you are likely to encounter them throughout the Middle East. It is clearly the best Beretta battle pistol made prior to the M92 and is the equal of any other single-column, all-steel, single-action-trigger-style 9x19mm pistol—except, of course, the SIG P210. Given the choice between this pistol and a stock P-35 with the pre-Mark III safety, I would pick the M1951.

The author testing the Egyptian-manufactured M1951 9x19mm.

Right view of Egyptian-made Beretta M1951 9x19mm (below).

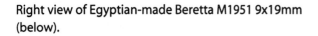

The M1951's unusual safety system allows the shooter to apply the safety by pushing in with the knuckle of his thumb. He can disengage the safety by pushing with the edge of the shooting finger knuckle. This is a very fast system, yet it avoids the problem of inadvertently bumping the safety off with car seats or belts, as happens with more typical safety systems. Until you realize how the system works, it seems awkward. But you soon appreciate its effectiveness (above).

The 50-foot test target show with the M1951.

Type 26 Revolver

9x22mmR
(9mm Japanese Revolver)

A review of all the Japanese-designed and manufactured handguns shows that the Japanese had no real handgun background. It also shows that they may have had good reason to prefer swords. To most Westerners, the idea of using a sword instead of a pistol must seem quite odd. If you use their handguns, however, you begin thinking that 30 inches or so of sharp steel may well be better than an 8/9mm pistol of questionable design and limited power.

The Type 26 revolver was allegedly designed for Japanese cavalry units, who were commonly deployed in China from 1894 through 1945. This is why the pistol is designed with no hammer spur and fires double-action only. Such a feature is generally safer from the back of a horse than a semi-auto single-action-type self-loader. In fact, having shot the Type 14 and Type 94 Nambu semiauto handguns, I must say that, except for the locking-

bolt problem, I would prefer a Type 26 to them.

The 9mm Japanese round is similar in power to the .38 Smith & Wesson, thus the recoil and blast were quite moderate.

The front sight is gray and rounded, and the rear sight is tapered and has a very small notch. As a result of the taper, the front and rear sights blended together, making it impossible to pick up the front sight rapidly for quick indexing. The front sight was totally lost on the cinema range; only the good grip feel made it possible to hit rapidly moving targets in poor light. Because of the small front and rear sights, and I found it impossible to hit 8-inch plates at 50 yards (although the ammunition may be partly to blame). My test group was 7 1/8 inches, pretty unremarkable. On the cinema range, the grip felt very good in my hand. The pistol's instinctive pointing ability is quite good, and despite the poor sights, I was able to score 90 percent on the cinema range.

SPECIFICATIONS

Name: Type 26 Revolver

Caliber: 9mm

Weight: 2 lbs.

Length: 9.4 in.

Feed: 6-rd., revolving cylinder

Operation: DA only, revolver

Sights: Front barleycorn; rear undercut notch

Muzzle velocity: 634 fps

Manufacturer: NA

Status: Obsolete

Recoil is low, allowing quick repeat shots, no doubt aided by the smooth trigger and fine double-action trigger pull. The weapon handles well and balances nicely. The double-action pull is quite good, and the extraction system is very positive. It is also fairly lightweight and really would not pose a burden to anyone.

Two last points of interest, one good and one bad. On the positive side, this pistol has a swinging sideplate, as does the Austrian Rast & Gasser. You simply swing it open to replace parts or clean the weapon. Most pistols make it very unwieldy for anyone other than a trained armorer to see the insides, and this results in many weapons never being cleaned properly. The swinging sideplate is a huge improvement over what is available on Webley, Smith & Wesson, and Colt revolvers even today.

Now for the bad news: this pistol has one feature that will ban it from the hips of all real handgun fighters—the lack of a bolt locking on the cylinder. What this does is allow the cylinder to rotate back to the chamber already fired. The result is that you pull the trigger and hit a case that has already been fired. This is an unacceptable position to be put in, and hence the weapon must be rejected. Given a choice of the Type 26 or a sword, I would pick the sword. How the Japanese cavalry must have envied the Chinese their Mauser pistols!

Left view of the Japanese Type 26 revolver and the 50-foot test target shot with it.

The Type 26 with action open. The ability to quickly open the entire action for full cleaning is an excellent feature for a combat arm.

The open action demonstrates how quickly the cases can be ejected.

A close-up of the hammer on the Type 26. The heavy hammer improves reliability, and the double-action-only design shows the clear intent of the designer to produce a combat revolver, not a target arm.

Nambu
(Papa)
8x21mm
(8mm Nambu)

This pistol was designed by Kitiro Nambu, a Japanese army officer, as a commercial pistol. It was manufactured in four variations at various plants. It was purchased by individual officers, and those in service were give an army inspection marks. Some were factory-stamped "Army Type"—probably as a sales ploy—but the only "army" order was some 500 that were purchased by Siam before World War I. Some models were also designated "Fourth-year-type naval pistols," and an officer's manual was printed for them.

When evaluating the Papa Nambu, you must recall that Japan had only joined the modern world in 1853, that it had no tradition of handgun use, and that the available European designs of the periods were Lugers, Dreyses, Bergmanns, and Mausers. The caliber selected was chosen at a time when it was not clear to designers that the small-bore concept applied to military rifles, with the

SPECIFICATIONS

Name: Nambu (Papa)

Caliber: 8mm

Weight: 1.9 lbs.

Length: 9 in.

Feed: 8-rd., in-line, detachable box mag.

Operation: Recoil; semiauto

Sights: Front barleycorn; rear tangent w/ notch

Muzzle velocity: 1,065 fps

Manufacturer: NA

Status: Obsolete

adoption of smokeless powder, did not necessarily work with handgun ammunition.

This pistol is well made of good materials. The sights allow quick indexing on the cinema range, due to their being high and obvious in semidarkness, but on the formal range such pyramid sights are harder to use. Obviously, the caliber is light, but certainly it is no worse than the 8mm Roth used in the Roth-Steyr 07, which I view as one of the top three World War I-era combat handguns.

The safety system leaves a lot to be desired; the weapon cannot be safely carried loaded with a round in the chamber. Instinctive/reactive shooting is more difficult, therefore, than with a Roth-Steyr 07. Still, it is no worse than a Luger, which is difficult to get into action quickly.

Accuracy on the formal range is surprisingly good. All of the Japanese weapons tested proved to be good performers in that regard. Perhaps my expectations were too low, and that caused me to be surprised. But the low recoil generated by the 8mm Nambu cartridge in a medium-weight weapon, coupled with a trigger system that is light if not crisp, permits accurate shooting.

On the cinema range, the angle of the grip, which is quite similar to that found on the Glock, allowed good instinctive shooting. Quick indexing was made easier by the high front sight (although a dab of white paint on the front and on each side of the rear blade would speed up pickup in poor light).

The low recoil of the 8mm Nambu cartridge and the low muzzle flash permit rapid repeat shots. Once in the hand, the weapon is as good as a Luger; it is only on reaction/instinctive draw and shoot courses that the design is so obviously weak.

The author firing the 8mm Nambu pistol.

The Papa Nambu was very expensive to produce, and most of the production machinery was destroyed in the 1923 earthquake. Further, the army wanted to adopt an official pistol that was cheaper. The Taisho Type 14 was easier to make and offered a more positive safety. If anything, it slowed up response time, but I suppose the Japanese soldier was more concerned about accidental shootings than instinctive response times.

Similarly, the quality of the Type 14, even at the beginning, was not as high as that of the Papa Nambu, which was virtually a hand-fitted, if not handmade, weapon. The most consistent mechanical

Right side of Nambu 8mm pistol.

problem with it was weakening striker springs, which resulted in misfires. As a result, the Papa Nambu is much less common than the Type 14. You may find one in remote places infrequently visited, and, if so, it will serve you as well as any other weapon of its era, such as a Brixia, Luger, or Dreyse. Used within limitations and with ammunition that is fresh, the Papa Nambu is an adequate combat handgun.

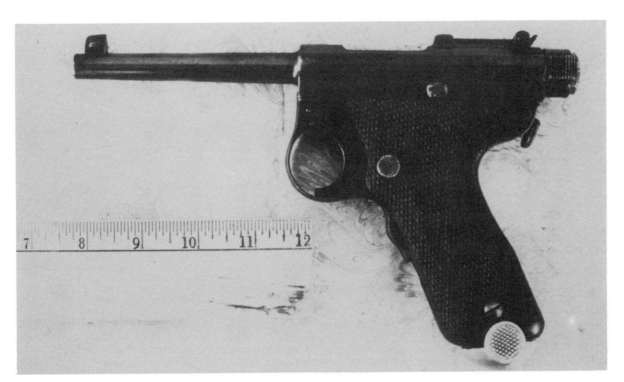

Nambu 8mm.

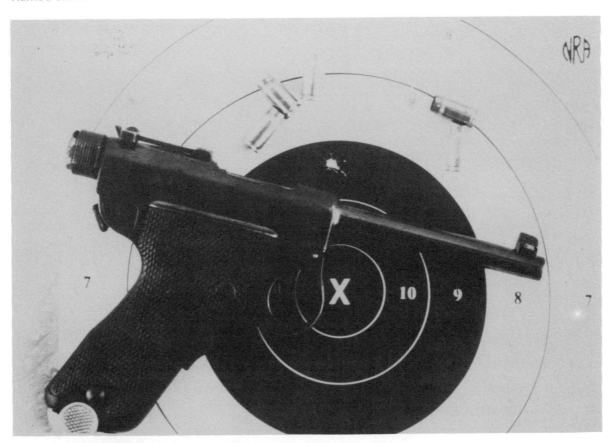

The 50-foot target fired with the 8mm Nambu.

Type B
"Baby" Nambu

7x19.7mm
(7mm Nambu)

As with the Papa, the so-called "Baby" Type B Nambu was designed as a commercial venture to be sold to Japanese officers, who were required to purchase their own sidearms. Much about the Baby Nambu is shrouded in mystery, and you frequently run across comments that the weapon was designed to be used by general officers only. I have never seen anything to confirm this notion, and I think people assume it because they are familiar with the U.S. practice of issuing Colt pocket models to ranking officers.

It seems likely that many Baby Nambus were used by high-ranking military officers, but I do not believe its use was restricted to them. It was available commercially, albeit in somewhat limited production. It also used a cartridge that was low in stopping power and unique to this weapon. Quite likely, the officer who purchased one did so from a desire to have a small

---

SPECIFICATIONS

Name: Type B "Baby" Nambu

Caliber: 7mm

Weight: 1.4 lbs.

Length: 6.75 in.

Feed: 7-rd., in-line, detachable box

Operation: Recoil, semiauto

Sights: Front barleycorn; rear V notch

Muzzle velocity: 1,050 fps

Manufacturer: NA

Status: Obsolete

Japanese-made pistol since European-made pistols were available then. I recently encountered a .25-caliber FN autoloader that was a documented captured weapon once carried by a Japanese soldier, for instance.

The 7mm Nambu pistol is certainly is not a "junker" by any means; it is well made and has good materials. Basically a miniature of its predecessor the Papa Nambu, it shares all of the Papa's design limitations and successes, including a good trigger and a front sight that is quick to index on the cinema range. The safety features of the Nambu 7mm leave much to be desired, but I suppose if the carrier left his pistol in the holster, withdrew it only when anticipating imminent danger, and then chambered a round, it would be adequate. Certainly the holster that was fitted to my pistol, which was standard for the weapon, would not lend itself to a speedy withdrawal.

The lack of a reactive safety might not be

worrisome, but the caliber certainly is lacking in power. Although it is true that with proper ammunition (such as that found in Soviet 5.45mm) a small-caliber weapon can be surprisingly effective, the 7mm Nambu ammunition does not have the benefit of a steel penetrator. To put it in perspective, the 7mm Nambu seems to be better than a .25 ACP and slightly less powerful than a .32 ACP. Despite the fact that it fires from a locked-barrel position, the 7mm Nambu does not have the power of .30 Luger or .30 Mauser cartridge. I do not know if the cartridge could be loaded to such a power level, but this would make it more useful. However, I do not know anyone who is willing to experiment with his 7mm Nambu to see whether the weapon could handle a more powerful load. Certainly, I am not willing to risk my Baby Nambu.

Factory-loaded 7mm Nambu ammunition is difficult to find, expensive since it is a collector's item, and of questionable quality since it is so old. So the only way this weapon could be tested would be to have the ammunition custom loaded. This was done by manufacturing cases, using established loads and custom lead bullets. Results were in line with what could be expected for a small-caliber, small-size pocket pistol. Because it is a lightweight weapon, recoil was surprisingly sharp, but not unpleasant.

After testing the weapon on the standard range, I tested it on the darkened police cinema range. I am certain that the only person in the world firing on a cinema range with a 7mm Nambu pistol that night was me! The lack of a good, safe way to carry the weapon slows up response time, but once in hand, it works well enough on rapid repeat shots and quick indexing.

Although it will never be a quick-reaction weapon or a "stopper," it is certainly more

powerful and of much better manufacture and materials than the .25 Beretta used by the Italian air force during the same time. I suppose to be fair about it, that is really how you must evaluate the 7mm Baby Nambu. It clearly is not a Glock 17, but it is as good as the FN and Beretta .25, smaller than the M1910 Mauser .25, and obviously more powerful than any .25 ACP pistol. Although the ammunition is difficult to get today, such concerns did not matter to the Japanese officer who could not get .25 ACP any more easily than 7mm Nambu.

The weapon feels good in the hand and indexes well on the cinema range; in fact, it is far superior to many European autoloaders of the period for those purposes. Formal accuracy is quite sufficient for its purposes.

Even in your most far-reaching journeys, you are not likely to come across the 7mm Baby Nambu and be forced to rely on it for your personal defense. However, if you did, you would still have an adequate weapon in your hand, and I for one would just as soon have it than a more traditional .25 ACP self-loader.

The author firing the 7mm Baby Nambu . The small pistol is almost lost in the author's size-9 hands.

Right side of Baby Nambu 7mm pistol.

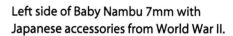

Left side of Baby Nambu 7mm with
Japanese accessories from World War II.

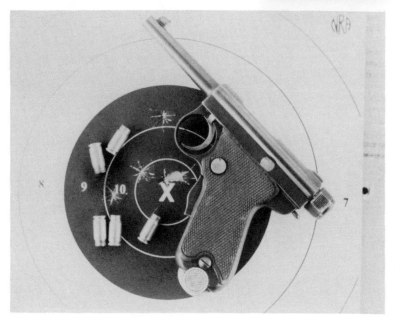

The 50-foot target fired with
the 7mm Baby Nambu.

Taisho Type 14
8x21mm
(8mm Nambu)

The Type 14 pistol is deprecated by many people, but it is still found all over the world. They turned up in the late 1960s and early 1970s in Vietnam, left over from the days of the Japanese occupation, and you can probably still find this weapon in odd corners of Southeast Asia.

Ammunition is a problem in 8mm Nambu because factory ammunition has not been manufactured for more than 40 years now, at least not in Japan. I tested with Mid-Way Arms reloads because the factory loads were unavailable, and I found them to be perfectly adequate so far as accuracy is concerned.

The first official-issue Japanese pistol, the Type 14 was designed by an ordnance board working from the Papa Nambu. It is really an unusual, well-made pistol of good materials.

I encountered no malfunctions of any type during the course of the test, which lasted

SPECIFICATIONS

Name: Taisho Type 14

Caliber: 8mm

Weight: 2 lbs.

Length: 9 in.

Feed: 8-rd., in-line, detachable box mag.

Operation: Recoil; semiauto

Sights: Front barleycorn; rear undercut notch

Muzzle velocity: 1,065 fps

Manufacturer: NA

Status: Obsolete

approximately 50 rounds. My test groups at 50 feet ran approximately 5 1/2 inches, again roughly twice that of the Model 19. I was so surprised at the size of the group, however, that I fired 10-shot groups and found that all 10 went into 6 inches. Clearly, the pistol is capable of acceptable combat accuracy.

The trigger pull on this Type 14, much like the one that I had 25 years ago, is very light. It is spongy, but I would be surprised if it weighs 2 1/2 pounds. It is very sudden and much too fine for military purposes.

The sights are very small. They are fine for formal target work, but not on the cinema range. On the latter, they get lost in the dark, where they tend to be gray. I could not see the back sight at all, and indexing ability was seriously impeded. The angle of the grip, however, is quite good, and that did allow some natural pointing ability. The trigger guard on the example that I tested was the "winter trigger guard," and that odd-size guard does not lend itself well to a firing stance with the left index finger over the trigger guard.

One of the pistol's worst features is its safety. It is very difficult to flip it off with one hand. You have to flip it through an arc of 180 degrees; no one in the world has a thumb long enough or strong enough to hit that safety and flip it off. So you are forced to use two hands to take it off or put it on, and, when using a normal two-handed hold, you can flip the safety off with your weak hand quite rapidly and fire the pistol with your other. On the positive side, the safety seems to be very certain. I cannot imagine anyone brushing it off by accident. However, the light trigger pull means you have to be very cautious about not putting your finger on the trigger until you are ready to fire.

Another mixed blessing on this particular pistol is its magazine disconnector, which in itself is desirable in a military pistol. But the spring that holds the magazine in place is so strong that you have to literally yank the magazine out with your spare hand. When you push the magazine release button, the magazine will drop just briefly, and then you have to pull it out; it does not drop from the weapon as it should. Possibly, this was designed to for the rigors of jungle warfare, where hitting the gun's bottom inadvertently could cause the magazine to drop into the mud. This design feature is commonly found on European military pistols, but it seems to be harder on the Nambu than on any other I have encountered.

Another interesting feature about this pistol is the magazine's short overall length, which keeps the trigger reach short, especially critical for short-fingered Asians. The cartridge angle is quite steep in comparison to that of a normal magazine; thus, you end up with a bottle-necked, fairly long cartridge being able to fit into a magazine that is quite short overall. At first, it would appear

that the steep angle might cause some feeding problems, but I encountered none during the testing of either this pistol or the Type 94, which also uses the same type of steep magazine angle. This feature is worth considering when you need a long, powerful cartridge but still want a short overall trigger reach for normal-to-small hands.

The caliber is nothing to write home about, but it is nearly as good as any 7.65 Luger. The people in Switzerland with their Type 29 7.65 Lugers are probably no better off than (and may not be as well off as) people in Japan with their 8mm Nambus. It is a moderate-velocity, small-caliber, lightweight projectile of round-nose design and, as a consequence, does not have much stopping power with conventional ammunition.

I believe that the Nambu is an underrated pistol. It is as good (or nearly so in my opinion) as any Luger pistol, and it has a better trigger pull than any Luger that I have ever seen. It should not be your first choice of a weapon, but it can be an effective combat pistol.

The author firing the 8mm Type 14 Taisho pistol.

Side view of the Type 14's magazine retention spring. The large depression helps with the withdrawal of the magazine even when wearing gloves.

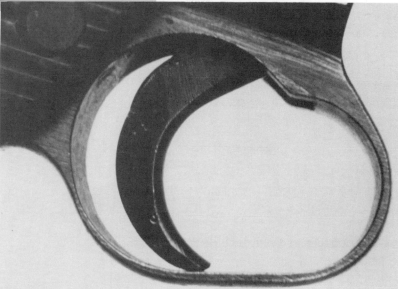

A close-up of the Type 14's enlarged trigger guard, an excellent feature.

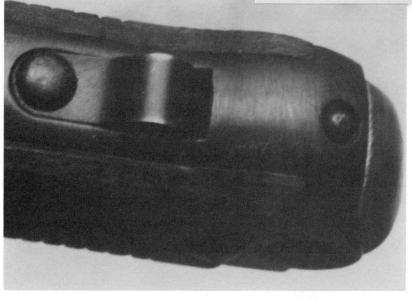

The magazine retention spring on this Type 14 prevents the accidental dumping of the magazine into the mud or earth. It also hinders a quick magazine release.

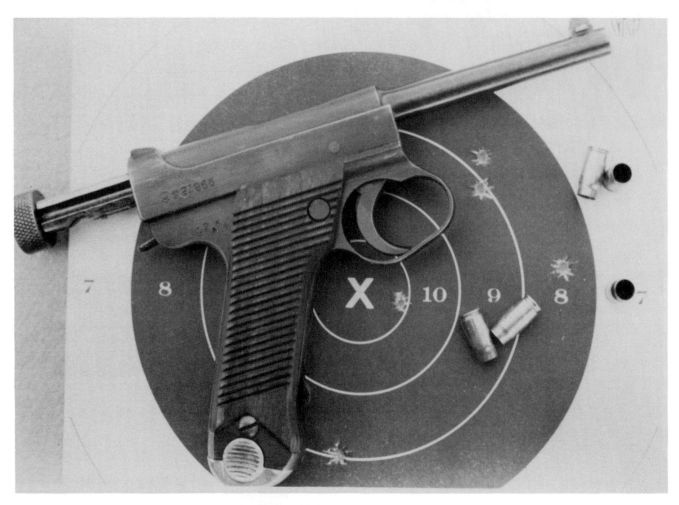

The 50-foot target fired with the Type 14 8mm Taisho.

Type 94
8x21mm
(8mm Nambu)

I think most people who disparage this pistol have never really shot it, or, after capturing it during World War II, they shot a few rounds at some tin cans when they weren't wearing any earplugs and then they wrote about it 30 years later. That is the only explanation I can give for the poor reports on it.

Although it is by no means a wonderful pistol, it is no worse than many European 7.65 pocket pistols that I tested. The sights were low, small, and hard to see, but they could be used for formal shooting if you worked at it. In other words, they were acceptable in good light. On the cinema range, the sights were much too shallow in the rear. The front sight was small, short, and dark, which made indexing difficult.

It did produce a lot of flash on the cinema range, and the recoil seemed quite high compared to the power of the cartridge. The cartridge has only 104-grain .30 caliber bullets at 900 fps

SPECIFICATIONS

Name: Type 94 Nambu

Caliber: 8mm

Weight: 1.7 lbs.

Length: 7.2 in.

Feed: 6-rd., in-line, detachable box mag.

Operation: Recoil; semiauto

Sights: Front barleycorn; rear square notch

Muzzle velocity: 1,000 fps

Manufacturer: NA

Status: Obsolete

velocity, so it should not kick that much. The pistol is light, but I do not believe that is the reason for the recoil. The real reason, I believe, is the grip design. The ropelike grip tended to shift in my hand, which makes control difficult and subsequent repeat shots much slower. This pistol kicked more than the Colt Government Model, in my hand at least.

There were no malfunctions of any type in more than 50 rounds, and 5-shot groups showed accuracy figures of 6 inches at 50 feet, with four of those rounds within 2 1/4 inches—not a shabby performance.

The safety in this particular pistol falls readily to hand, much like the Colt Government Model safety, and thus you can carry this pistol cocked and locked. There is a detent that goes into effect after you push it off, which makes it difficult to flip it back on with your thumb alone. But, of course, it is more important to be able to flip it off readily than it is to engage it quickly. If you worked on it a bit, you might be able to flip it back off just as easily, but I like to test them as I find them. The safety location, however, was quite convenient and did permit rapid disengaging of the safety and sighting in on targets.

Left side of Type 94 8mm Nambu pistol.

Toward the end of the war, quality control deteriorated dramatically in Japanese arms. In late-production Type 94 pistols, specimens that will fire out of battery are not uncommon. Although the 8mm Nambu is not a particularly "hot" round, when fired in a Type 94 pistol that is out of battery and thus without the breech locked, it is patently unsafe. Type 94 pistols

Right view of Type 94 8mm Nambu with the 50-foot test target.

that show rough workmanship should be checked before firing.

Unlike with the Type 14 pistol, which I also tested, the same loads in this pistol produce quite a bit of flash (even considering the barrel length differential). I am uncertain why this is the case, since I used the same ammunition.

Overall, the Type 94 Nambu pistol is not a dream pistol, but it is not nearly as bad as we

have been led to believe. I would rather have a Type 94 Nambu 8mm pistol than any Dreyse 7.65 pistol manufactured in the world. I think the pistol is better than the Dreyse: it is easier to disengage the safety, faster on the repeat shots, and has more power. If you are given a choice between some European pocket pistols in 7.65 and the Type 94, pick the latter, as long as you can get ammunition for it.

Detail of the safety on the Type 94. When in operation, the safety fell under the thumb, much like the Colt Government Model, and allowed the shooter to quickly engage and disengage. This safety system is far superior to those found on Japanese self-loaders.

A close-up of the much-hated trigger bar that bothers so many shooters. Yet, when the weapon was on safe and properly fitted, I could not get any pistol to accidentally fire when the trigger bar was pushed in.

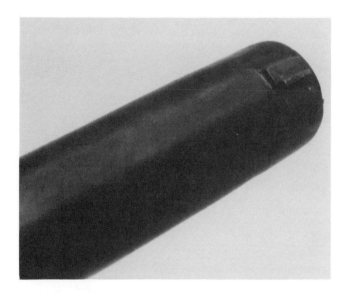

The front sight on the Type 94 allows good indexing, although it would be much better if painted white.

The rear sight on the Type 94 is low and dark but of adequate width.

Rast & Gasser "Montenegrin"

11x35.8mmR
(11mm Gasser)

This is a very interesting weapon for a number of reasons. First, King Nicholas of Montenegro (who reigned from 1910 to 1918) was rumored to have decreed that all adult males must purchase and carry such a weapon. I don't know if this is true, but certainly the philosophy was correct, in my estimation. If true, it is only unfortunate that the king did not specify the Colt Government Model; his subjects would have had a better piece of equipment.

As it was, the Model 1870 Rast & Gasser revolver purchased as surplus from Austria-Hungary became virtually a part of the Montenegrins' wearing apparel or costume. A wonderful picture of the king visiting with Britain's Field Marshall Haig during World War I shows this clearly (p. 259), and travel books of the pre-1914 period comment on the fact that the populace carried these large pistols in sashes around their waists.

SPECIFICATIONS

Name: Rast & Gasser "Montenegrin"

Caliber: 11mm

Weight: NA

Length: NA

Feed: Revolver

Operation: DA

Sights: Front blade; rear notch

Muzzle velocity: NA

Manufacturer: Gasser and various Belgium manufacturers

Status: Obsolete

These pistols are quite large and heavy—certainly not concealment pieces! After the Austro-Hungarian surplus ran out, they were made mainly in Belgium by a variety of manufacturers and were widely available in a variety of finishes prior to 1914.

Since factory-fresh ammunition was unavailable for my test, I had to have it made. My ammunition maker turned out excellent ammunition for the other pistols I tested, but the ammunition for the Rast & Gasser was far from perfect, and, as a result, accuracy suffered. I would have liked to have had the barrel slugged, but I was unable to keep the weapon long enough to do it. He built the ammunition to the proper specifications, but weapons such as this are frequently outside of specifications. The crown on the barrel also looked suspect to me, and if it had been mine, I would have had it recrowned. Otherwise, this inaccuracy did not affect the performance of the weapon or its use on the cinema range.

The trigger pull on the example tested was simply awful. It not only made it difficult to shoot on the formal range, it also hindered fast, accurate work on the cinema range. The barley-corn front sight is fast to use, although the rear sight is very small and shallow, thus complicating elevation retention and indexing.

The 9 1/2-inch barrel throws the weapon out of balance and makes proper presentation from a holster difficult. I suppose that is why the natives wore them in sashes cross-draw style.

The grips are correctly shaped to be narrow at the bottom yet not wide at the top. In fact, this is probably the pistol's best feature; the grip feels good in my hand without any type of adaptor.

Although the Rast & Gasser is not in the Webley class, a knowledge of it is still useful. They remain plentiful in the Balkan region.

Part of this weapon's interest for me involves its use by an acquaintance of mine, an ex-SAS member assigned duties in MI5. Once, while operating in the Middle East, he was totally without a weapon. He finally managed to get his hands on a Montenegrin 11mm

The trigger pull on the Rast & Gasser Montenegrin revolver was quite heavy, as can be discerned from the expression on the shooter's face.

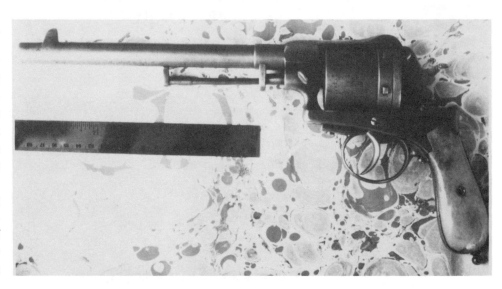

Left view of Rast & Gasser Montenegrin 11x35.8mm.

revolver, with which he was able to both defend himself and accomplish his mission. This reiterates quite clearly that a working knowledge of a variety of weapons may serve you well at some distant, desperate time.

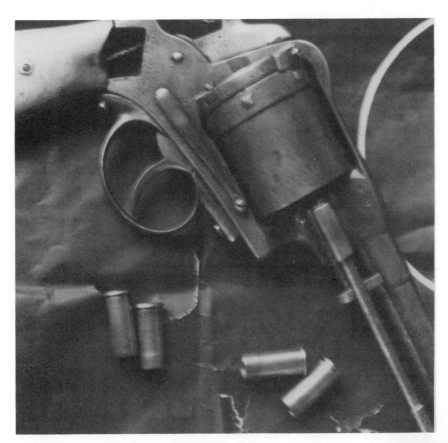

Heavy trigger pull and ammunition that may have been faulty resulted in very large groups in the 50-foot test.

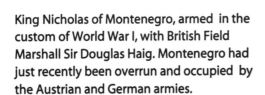

King Nicholas of Montenegro, armed in the custom of World War I, with British Field Marshall Sir Douglas Haig. Montenegro had just recently been overrun and occupied by the Austrian and German armies.

M1894 Ordnance Revolver

9.4x20.7mmR
(9.4mm Dutch Revolver)

This is another one of those revolvers adopted in the last two decades of the nineteenth century when the European powers went to smaller-caliber rifles and pistols with the universal acceptance of smokeless powder. Unlike many countries, however, the Netherlands at least used a 9.4mm bullet rather than the very small 7.5, 7.62, or 8mm, as was common in Scandinavian countries, Russia, and France. Perhaps their experiences in colonial Southeast Asia with hostile tribes had convinced them that smaller calibers would not work. Even so, the 9.4 Dutch is only equal to the .38 Smith & Wesson or .38 Long cartridge—not exactly a powerhouse.

The heavy double-action trigger pull forced me to place my arm at an angle to get enough leverage after six shots to fire the pistol. An examination of the weapon's insides showed that all parts were numbered properly, so I must con-

---

SPECIFICATIONS

Name: M1894 Ordnance Revolver

Caliber: 9.4mm

Weight: NA

Length: NA

Feed: Revolver

Operation: DA

Sights: Front bead; rear notch

Muzzle velocity: NA

Manufacturer: Artillerie Inrichtungen

Status: Obsolete

---

clude it was designed to be heavy, perhaps as a safety feature. But this heavy pull caused it to be very slow on double-action strings. The trigger surface is smooth, which normally helps rapid double-action work, but the pull limited its usefulness. However, had it been grooved or checkered, it would have surely harmed my finger.

Its reloading technique is similar to that found on a Colt SAA; the ejector rod is attached to the ejector housing and spring loaded. Each chamber has to be indexed to accomplish ejection. The small cutout on the frame at the rear of the cylinder has to be carefully aligned in order to eject the cases, which slowed up the procedure.

The rear sight was too low and shallow for rapid pickup. The front sight was quite tall and has a bead built into it, which would normally help pickup on the darkened cinema range. However, due to the angle on it toward the muzzle, light caused it to fade out. If it was painted white, it would show up quicker and be easier to use.

For its period, the Ordnance Revolver is not a bad example of a military revolver—although loading is much slower than with the Webley. However, a better comparison might be with the Colt M1878 in .38 or Lighting Model. When compared to them (remember the Dutch pistol was adopted in 1874), it fares quite well.

Perhaps the most interesting use of this pistol was its (or a similar civilian model's) use by the Dutch police before World War I. They would pattern load it with two blank rounds, a tear gas round, a shot shell, then two ball rounds. This is not nearly enough deadly-force application for this old ex-U.S. marshal. However, I guess times were a lot different then.

The author testing the M1894 Ordnance Revolver.

Left view of the M1894 Ordnance Revolver.

The M1894 Ordnance Revolver with original ammunition.

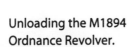

Unloading the M1894 Ordnance Revolver.

The 50-foot test target for the Dutch M1873 Ordnance Revolver.

M1914
.45 ACP

This pistol is basically a Norwegian-manufactured copy of the Colt M1911. The only real differences are that the slide stop is lower than on the Colt, thereby allowing quicker release of the slide, and the grip is cut out to accommodate the lower slide stop. Otherwise, it is a standard M1911.

The Norwegian authorities either went crazy with the number stamps or the manufacturer was concerned about the interchangeability of parts because several numbers appear on the weapon. Obviously, this makes no difference to the shooter, and it warms the hearts of the collector.

The Norwegian M1914 shoots exactly the same as the M1911. The short tang causes some hammer bite if not held correctly, just as with the

---

SPECIFICATIONS

Name: M1914 .45 ACP

Caliber: .45 ACP

Weight: 2.4

Length: 8.62 in.

Feed: 7-rd., in-line, detachable box mag.

Operation: Recoil; semiauto

Sights: Front blade; rear square notch

Muzzle velocity: 830 fps

Manufacturer: Arsenal, Kongsberg

Status: Out of production; limited use

M1911. The Norwegians made these pistols in small quantities up to World War II, and a small number were manufactured for the German occupation forces.

Possibly the best thing about the M1914 for the practical shooter is that these pistols can tell no tales. With limited record keeping by the Norwegian military, a foreign language not commonly encountered, and the German occupation, it is unlikely that any government authorities can trace the lineage of any of these weapons. That's always a positive thing.

The Norwegian army picked a good handgun to replace the 7.5 Nagant revolver. The Colt automatic was a giant step forward in power, reloading ease, and maintenance. At that time time, there was not an overabundance of suitable self-loading pistols.

The author testing the Norwegian M1914.

The 50-foot test target for the Norwegian M1914.

Radom Model 35

9x19mm
(9mm Para)

The Polish government adopted the Radom pistol in 1935 and manufactured it for the Polish army through the beginning of World War II. After Germany conquered Poland, the Germans continued to produce this pistol in Poland for their army; some were also assembled by Steyr in the last days of the war. It has not been produced since 1944.

Rumors have run rampant about the hammer-drop safety on this particular pistol. It has been said that the pistols were sabotaged by slave workers, so that when the hammer was pushed down, it would drop and fire. Perhaps some examples had overlong firing pins, but the example tested certainly did not have that problem.

When evaluating this pistol, you have to remember where it came from and when it was adopted. Pre-war Poland had the most cavalry units around, and the Radom was designed to be fired one-handed from a horse. The pistol has

---

SPECIFICATIONS

Name: Radom Model 35

Caliber: 9mm Parabellum

Weight: 2.3 lbs.

Length: 7.8 in.

Feed: 8-rd., in-line, detachable box mag.

Operation: Recoil; semiauto

Sights: Front blade; rear U notch

Muzzle velocity: 1,150 fps

Manufacturer: Fabryka Broni, Radom

Status: Obsolete

---

only a grip safety, and the lever on the side that looks like a safety is merely a hammer drop. Pushing this lever down drops the hammer; releasing it permits it to spring back into the original position.

The weapon that I tested had a poor trigger pull, and that caused larger groups— 4 1/4-inch groups—than otherwise would be found.

It also had poor sights, which decreased accuracy. They are too small with the "U"-shape rear notch and a very small front, making them difficult to see for normal target work and even worse on the cinema range. Its performance on the cinema range was saved only because its feel was so close to the Colt Government Model, with which I am familiar, that I was able to respond instinctively.

The fact that no cock-and-lock safety exists means that you have some choices to make about carrying this weapon. You can carry it with the hammer down and then try to cock it, which is difficult to do because of the burr hammer and the fact that the side of the slide covers quite a bit of the hammer. Alternatively, you can carry it with the chamber empty and hammer down, which requires the use of two hands to get the weapon to operate—an unacceptable option for me. Or you can carry it cocked and trust the grip safety. The grip safety did keep the weapon from going off, but I would not trust it. Folklore has it that the Polish cavalry units were supposed to carry it with the hammer down on a loaded chamber, which is what the hammer-drop safety was for. Then when a soldier pulled his weapon, he was supposed to brush it against his saddle so that the burr hammer would cock, and it would be fired one-handed. That strikes me as an extremely dangerous practice. One always wonders exactly how many horses were shot as a consequence of this method.

The grips flare at the bottom, and this breaks your grip because, obviously, grips should taper, not flare, at the bottom—much like the hand getting smaller toward the bottom of your fist rather than bigger.

The Radom is a reliable (though somewhat heavy) pistol because it is all-steel and has a single-column magazine. The magazine release has a button on the side much like the Colt Government Model from which it was designed, and it did drop the magazine out reliably. However, its safety features inhibited getting it into action quickly and safely.

In summary, although the Radom can be quite accurate (I found I was able to hit a chest-size metal plate at 100 yards with a two-hand hold without any problems), its poor trigger pull, too-small and difficult-to-index sights, and problematic safety make it unacceptable for military purposes. Better choices exist if you are looking for a single-column, single-action 9mm pistol.

The author testing the Radom M35.

A close-up of the hammer area of the Radom M35. Polish troops were taught to cock the weapon by brushing it against their pant legs!

Right side of the Radom M35 and the 50-foot target shot with it.

Savage M1907
7.62x17mmSR
(7.65mm Browning, .32 ACP)

During World War I, this pistol was a substitute standard weapon for French forces and the standard duty weapon of the Portuguese army. A friend reports encountering one in Portuguese East Africa during the 1970s. It was also a popular pocket pistol with many travelers before the war and is likely to be encountered all over the world even today.

"Bat" Masterson thought very highly of the Savage and claimed it was superior to the revolvers of his days. Certainly the Savage, with 10 shots, had the highest magazine capacity of any normal-size .32 autoloader. Despite the wide magazine, however, its grip is not too broad because of the cartridges it contains and the grip design.

The magazine release is rather hard to push in, and unless this is done perfectly, the magazine will not fall out. Although the 10-shot capacity limits the need for a rapid magazine change, the maga-

SPECIFICATIONS

Name: Savage M1907

Caliber: 7.65mm (.32 ACP)

Weight: 1.2 lbs.

Length: 6 1/2 in.

Feed: Staggered magazine, 10 shots

Operation: Delayed blowback

Sights: Front blade; rear notch

Muzzle velocity: NA

Manufacturer: Savage

Status: Obsolete

zine release is still a weakness. Additionally, the Savage has no hold-open device, so you cannot tell when it is empty.

The safety is easy to disengage, and you can also feel easily with your thumb if the safety is in place. However, I would not recommend that the weapon be carried with a round in the chamber and cocked, because the striker rests on the primer. With a slip of the thumb, which could happen easily with a burr hammer and wet hands, the weapon could fire.

The sights on this pistol are very small and dark, so it was quite hard to index the weapon properly on the cinema range. The rear sight is small and shallow, making triangulation difficult. Even on the formal target range, the sights were hard to see because of the narrow notch and shallow depth, and I am certain they contributed to the larger-than-average group of 4 1/2 inches that I fired.

The Savage M1907 is an old design, but one

that is unusual in its high magazine capacity and its elegant appearance. If you doubt that, compare it to a Ruby .32, which was also used by the French during the period. However, it never was the military weapon that the CZ 38 is even though it served longer and over a greater geographic area than the CZ 38 or similar weapons.

The author testing the Savage pistol.

The Savage 32 was one of the first autoloaders to use a double-column detachable magazine in a pocket pistol.

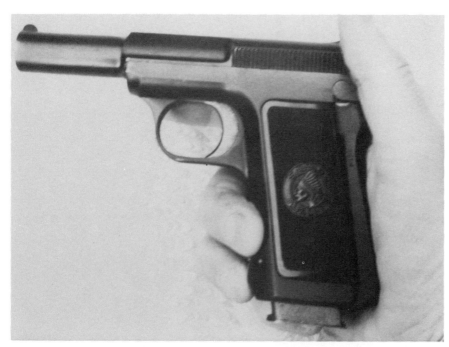

Carrying the Savage cocked and locked requires you to disconnect the safety, as illustrated here by using the thumb.

The 50-foot test target for the Savage.

Astra 400

9x23mm
(9mm Largo, 9mm Bergmann-Bayard)

This common-place pistol has some unique features, the main one being its ability to take a wide variety of ammunition. It is designed to take the 9mm Largo round, but in addition it also shoots .38 Super, .38 Auto, 9mm Parabellum, .380, 9mm Steyr, and just about any other 9mm type rounds it will chamber. Depending on manufacture, some of the rounds may not work perfectly. For instance, I found that Winchester full-metal-jacketed 9x19mm tended to drop into the chamber too far, and the firing pin would not properly dent the primer. On the other hand, 9x19mm Fiocchi ammunition worked just fine. So, accommodating the large variety of cartridges that you might find in all sorts of odd corners of the world is an advantage.

Also of course, this weapon is a type of pistol that seems to pop up without too much in the way of background, which is a big advantage. It is an enclosed hammer design with a heel-butt magazine release, and most such pistols generally give me bad

SPECIFICATIONS

Name: Astra 400

Caliber: 9mm Largo

Weight: 2.1 lbs.

Length: 8.7 in.

Feed: 8-rd., in-line, detachable box mag.

Operation: Blowback; semiauto

Sights: Front blade; rear square notch

Muzzle velocity: 1,210 fps

Manufacturer: Unceta & Co.

Status: Obsolete

vibrations because I view them as unsafe. But the safety in this pistol is simple to disengage, and if the grips were just a little thinner, it would be even faster and easier to use with the weak hand. The heel-butt-magazine release is a problem because there is a tendency to push these releases off when you are in your car seat as you tend to push the button backward and disengage the magazine.

This is a blow-back-only pistol like the Astra 600; as a consequence, the spring on it is quite stiff and requires a bit of muscle to pull the slide to the rear and cycle the weapon. However, it worked just fine without any apparent pressure indications. Recoil seems to be quite high for a 9mm pistol, again due to the blow-back-only design.

Accuracy was acceptable; I recorded 3 5/8-inch groups. The sights are somewhat hard to see, and the slide does not lock open at the last round fired, which is too bad. Grip angle is good, and I was able to get on the targets rapidly in the cinema range. The sights for formal target work are too small and are hard to pick up rapidly. Interestingly enough, however, on the cinema range, the sights can be picked up surprisingly easily, and because they were on the end of this tapered-type barrel, I could index rapidly on the target. I attribute that to the fact that the barrel is long and cylindrical; thus, I could pick up the end of the tube and triangulate rapidly.

All in all, I was quite pleased with the Astra pistols. I had viewed them as very junky pistols but found instead that they are quite satisfactory, given the caliber and weight limitations. In fact, based on stock weapons, I would rate the Astra quite a bit higher than many pistols that have a much wider following, such as the P-35 Browning.

The author testing the Astra 400 (above).

Right side of the Astra 400 (below).

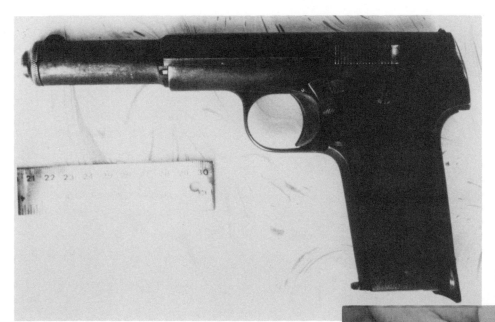

Left side of the Astra 400.

The blow-black action uses strong springs to keep the action closed rather than the traditional locking system. This results in a heavy pull being required to retract the slide.

The Astra 400 9x23mm with the 50-foot test target shot with it.

Astra 600

9x19mm
(9mm Para)

Whether the 600, 900, or the 400, Astra pistols are always surprising when you test them. On the formal target range, I was able to achieve 3 1/4-inch groups, which I consider outstanding given the weapon's poor sights and 12-pound trigger pull.

Recoil on these pistols is quite heavy because they are blowback only. Generally, 9 x 19mm pistols are not blow-back pistols. Because it uses a heavy spring, the Astra is still safe.

The magazine on the 600 I shot required two hands to remove, although the 400 that I tested did not have this problem. The safety on the 600 releases very rapidly, and on the cinema range it was quite a fast pistol. The grips are a little thick, but they could be thinned because there is more than an adequate amount of material for my size-9 hands. If they were thinned, the safety would then fall even more readily to hand (or thumb).

SPECIFICATIONS

Name: Astra 600

Caliber: 9mm

Weight: 2.2 lbs.

Length: 8.7 in.

Feed: In-line box magazine

Operation: Blowback

Sights: Blade/notch

Muzzle velocity: NA

Manufacturer: Astra

Status: Obsolete

Commonly, Astra pistols are thought to be junk, and many people shun them, choosing instead the Luger, Radom, or Beretta pistols. If I had the choice between an Astra 400 or 600 pistol and any Luger that has ever been made, my first choice would definitely be the Astra. Although the Astra may not look like a good pistol, on the cinema range or as combat pistol, it performs quite satisfactorily. The sights are small, too thin, and need to be painted a contrasting color, but their mount on the end of the tubular barrel enhances indexing.

Some of them were sold to the German army during World War II, and Astra continued to produce them after the war. They are common throughout the world, and it is likely that you will encounter these in many situations. Knowing the Astra's strengths could come in handy: the hammer design makes it easy to carry concealed, it is free from sharp edges, the safety is out of the way, and it is not likely that it would disengage accidentally.

Astra 900

7.62x25mm
(7.63 Mauser, .30 Mauser)

The Astra 900 is much like the Mauser M96: they are both 7.63mm pistols that are awkward to shoot on the formal target range. Their sights are difficult to see, their triggers are difficult to control, and their hammers fall with a heavy thud. If you fired them only on the formal range, you would wonder why anyone liked these pistols. Yet they were very popular in their time. In fact, only when the cost got prohibitive and their manufacture ended did they go out of fashion.

But unlike the Broomhandle Mauser, the Astra has some pins and screws in its construction, and its interior is carefully finished. The Astra was the first to bring out a selectable-fire weapon and make detachable magazines common. In fact, Spanish pistols of this type had features superior to the original German designs. The Spanish Astra Model "F," with its rate reduction in the grip, resulted in a pistol with a rate of

SPECIFICATIONS

Name: Astra 900

Caliber: 7.63 Mauser (.30 Mauser)

Weight: 2.7 lbs.

Length: 12 in.

Feed: Fixed box

Operation: Short recoil

Sights: Front blade; rear adj. tangent

Muzzle velocity: 1420 fps

Manufacturer: Unceta & Co.

Status: Obsolete

fire in the 350 RPM range—far superior to the 900 RPM of the Mauser 712 selectable-fire weapon.

It is unfortunate that two things conspired to limit the Spanish manufacture of these weapons. First, Spanish firearms were frequently thought to be made of soft metal and, therefore, were assumed to wear out faster than necessary and perhaps even be unsafe. Even today, that is still a problem. Left to its own devices, Star will make soft pistols. This is the reason that early Star PD 45s were rejected by the importer. Watched carefully, the Spanish can put out a good, strong, hard pistol. Why this should be, given the history of swordmaking in Spain I really don't know. Perhaps it is the lack of sophisticated metallurgy or the practice of paying for piecework (softer weapons can be made faster). The Astra 900, like many Spanish pistols, came on the scene somewhat suspect as to its strength.

Second, during the Civil War in Spain, weapon

production was taken over for local consumption. Once its war was over, Spain had associated itself with Hitler and become involved in another war; by the time World War II was over, the world was awash in low-cost surplus weapons of a more modern design.

As with the Mauser M96, the Astra 900 is an effective fighting weapon on the cinema range. The long barrel and high sight allow for rapid indexing in the dark. The light recoil permits rapid shots, while avoiding shifting of the weapon in the hand. As with my comments on the Mauser M96, this is a combat weapon; it is not a dilettante's paper punch. For someone who cannot carry a rifle or SMG, the Astra 900 (especially the detachable-magazine type with butt stock available) becomes a very serious fighting tool.

The author testing the Astra 900 .30 Mauser.

Right side of the Astra 900 .30 Mauser.

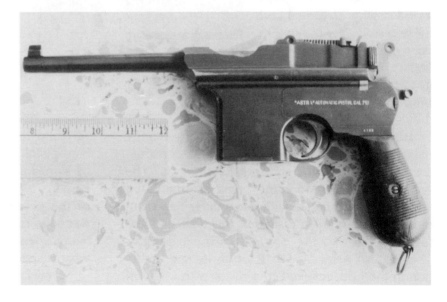

Left side of the Astra 900 .30 Mauser.

Star Model B

9x19mm
(9mm Para)

The Star Model B is a Colt Government Model-style pistol manufactured in 9x19mm. In Spain, the same pistol is available in 9.23mm (9mm Largo), .38 Super, 7.63mm Mauser, and .45 ACP. Star pistols all follow traditional designs and are well made. Although they always look good, occasional batches of Stars will come through on the commercial scene that are not as well made as desired, mainly due to poor quality control over the steel specifications. Sometimes made out of soft steel during times of crisis, they wear rapidly, the parts get out of alignment, and functioning suffers. If the steel is good, however, the pistols are fine weapons. If you have an ex-military or police Star, you can be fairly certain that the steel is good because the weapon would not have been accepted.

The Star Model B has all the good and bad points of a Colt Government Model. It does offer

SPECIFICATIONS

Name: Star Model B

Caliber: 9mm Parabellum

Weight: 2.3 lbs.

Length: 8.71 in.

Feed: 8-rd., detachable box mag.

Operation: Short recoil; semiauto

Sights: Front blade; rear notch

Muzzle velocity: 1,300 fps

Manufacturer: Star Bonifacio Echeverria SA

Status: In service

a magazine safety, which is a good feature on a military weapon, and the safety locks the hammer rather than the sear, also a design improvement. The weapon is made out of steel, and since it weighs as much as a Colt Government Model, the shooter (if firing a cartridge with less recoil) is able to fire repeat shots faster. On the cinema range, my performance was quite good, but since I was firing about 500 rounds a week through a Colt Government Model at the time, I do not think an impartial evaluation exists. The sights should be painted white for faster pickup under darkened range conditions.

The Star MMS (a 7.63x25mm version) is also available with a butt stock and extended magazine. For someone not equipped with a rifle, this butt-stock-equipped pistol might be just the ticket, especially since a 7.63mm pistol with proper bullets will penetrate ballistic vests better than more traditionally effective cartridges like the .45 ACP.

Buttstocks on pistols do not make the make the well-trained shooter more accurate on a formal range, but they do make it easier for the beginning shooter to use a handgun at long range; they also add steadiness for the cold and out-of-breath expert or inexpert shooter. There should be a place in the military arsenal for butt stock pistols, perhaps not a single-action, single-column Star but some more modern butt stock pistol.

The Star Model B was used by the German forces during World War II as a substitute standard pistol. Because Franco supported Hitler, the Star in 9mm Largo was available. German ordnance officers contracted for a modified model, and thousands made the trip over the mountains to France. Interestingly enough, according to a French publication I read recently, two of them were used at the Paris Gestapo Headquarters in 1944.

After the war, the Germans who started their new paramilitary Border Police adopted the Star for their officers, and many of the Star Model B pistols found in the United States today are ex-German border patrol guns. Since few saw much use, their condition is usually excellent.

However, it was in the hands of the South African military that the Star Model B's finest hour came. Prior to the embargo of weapons in 1964, South Africa bought the Star Model B to replace an aging stock of Webley and Enfield revolvers. Today of course, South Africa produces a home modification of the Beretta M92, but the Star Model B remains in inventory.

The best example of this military handgun in combat that I have been able to document involves a South African officer and took place during the Angolan War. Fighting against Cuban troops who were assisting the Angolan Communist forces, a South African commander in charge of an armored car had the tread of his vehicle knocked off. Cuban forces attempted to close in on the vehicle with buckets of gasoline. At first, the South Africans used the vehicle's light machine guns to drive away their attackers. When they ran out of ammunition, they turned to the submachine guns that came with the vehicle. Although many Cubans fell to the accurate and deadly SMG fire, the South Africans soon ran out of ammunition again. Reduced to using his Star Model B with standard 9x19mm full-metal-jacket ammunition, this South African downed Cuban after Cuban as they continued to attack his vehicle. He changed magazines and continued to fire out the slits of the vehicle. More Cubans dropped. When he was finally done, he had killed 34 Cubans with his Star Model B alone—truly an outstanding performance. More common (though by no means frequent) are the six to eight enemy soldiers killed with the military handgun.

During World War I, a series of Victoria Crosses were awarded for actions involving the revolver or pistol. The South Africans also recognized the feat of the South African who held off the Cubans and awarded him the highest award for valor available in South Africa. This incident illustrates that the Star Model B is a military handgun of high standing. If you can get a Model B, remember that you are in good company and that your equipment is capable of truly legendary performance.

The author testing the Star Model B 9x19mm.

Right side of the Star Model B
9x19mm.

Left side of the Star Model B with
the 50-foot target it shot.

The author test-firing the
Star MMS with stock affixed.

Right view of Star MMS 7.63x25mm (above).

The 50-foot test target shot with the stockless Star MMS (right).

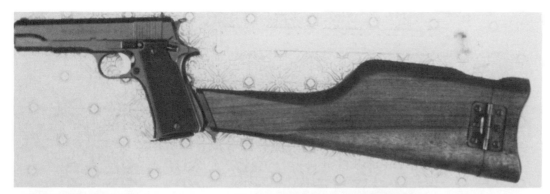

A Star MMS with stock affixed. This makes a neat package.

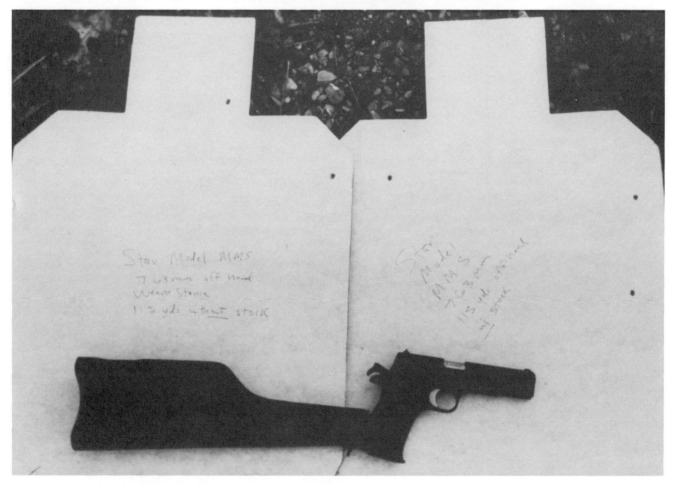

Two targets shot with the Star MMS at 115 yards. The group on the left was fired without a stock, while the one on the right was shot with a stock. Although the groups are basically the same size, firing with the stock was easier. If the shooter was tired, out-of-breath, or inexperienced, the results would likely be more dramatic in favor of using the stock. The more expert the shooter, the less useful the stock is, in the author's opinion.

M1887 Nagant

7.5x22.4mmR
(7.5 Swedish Revolver)

This revolver was also designed when smokeless powder was just beginning to become popular and the designers had not figured out that you could reduce the bore and still keep lethality high with rifles, but not with handguns (at least not those using the conventional projectiles available during this time).

Most of these revolvers are in excellent condition since all they did is keep the peace, not fight the war. Even though they are well made, a major drawback is their lack of power: the 7.5 Swedish round is little better than the standard .32 Smith & Wesson Long cartridge. Still, since these weapons are frequently considered antiques and many people do not realize that commonly available .32 Smith & Wesson Long ammunition can be used, such weapons are sometimes available when others are not.

On the formal target range, the triangular front sight was hard to use, but on the cinema range it indexed quite well. The double-action pull was heavy, although the trigger placement was very good for my fingers. The grip was fine for rapid fire and I encountered no shifting, but the recoil of the loads was very low. While the rear sight was impossible to use on the cinema range, the good grip angle and the ability to quickly index off the front sight yielded good results on the cinema range.

The weapon is lightweight, and its design makes clear that it is a safe, durable pistol. Reloading is slower than with the Webley top-break or Smith & Wesson/Colt simultaneous ejection because the Nagant's reloading mechanism consists of an ejector rod to poke the empties out one at a time. You twist the ejector rod housing to the side, pull the rod down each cylinder hold, and eject the empties. It is a slow procedure.

In summary, this pistol is adequate on the cinema range, not so good on the formal range, and

SPECIFICATIONS

Name: M1887 Nagant

Caliber: 7.5mm

Weight: NA

Length: NA

Feed: Revolver

Operation: DA

Sights: Blade/notch

Muzzle velocity: NA

Manufacturer: Husqvarna

Status: Obsolete

in a caliber that is too light for military or police use. Any caliber that is only suitable for head shots on rabbits or squirrels is not a weapon for me to pack when going in harm's way. The low recoil did allow for quick double-action shooting, but the trigger pull on the example tested was so heavy that it took some practice to learn how to avoid pulling the weapon to the side to compensate for the heavy trigger.

I am certain that Swedish soldiers were happy when the military adopted the 9mm Browning 07 pistol, and you, likewise, would be better off with that pistol as well, if given a choice.

The author test-firing the Nagant M1887 (above).

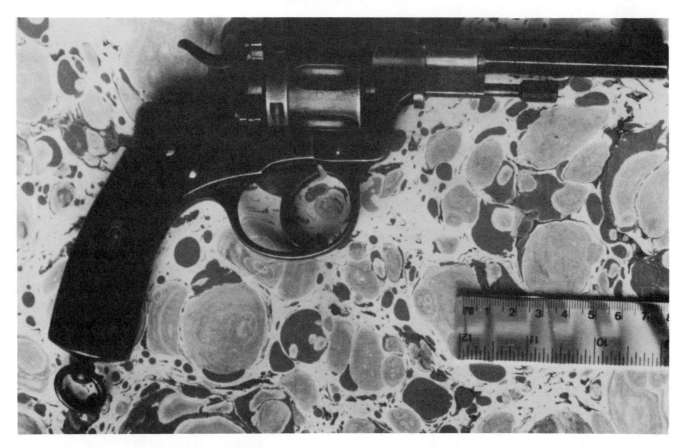

Right side of the Nagant M1887 7.5mm (below).

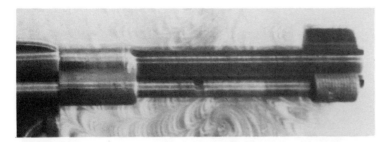

Emptying the cylinder on the M1887 requires that the ejection rod be shifted from under the barrel and them manually pushed back and forth.

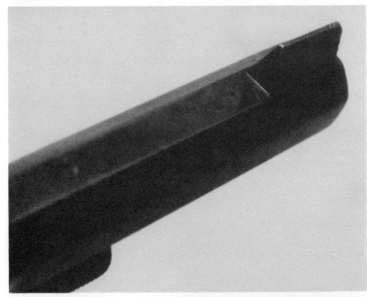

The high barleycorn sight on the M1887 Nagant improves indexing in poor light, but makes it difficult to hold elevation on the formal range.

The 50-foot target shot with the M1887 Nagant.

M1907 Swedish (Browning M1903)

9x20mmSR
(9mm Browning Long)

The M1907 Swedish self-loading pistol tested was a licensed copy of the Model 1903 Browning. It was one of the first semiautomatic pistols commonly adopted for military purposes and was available in the 9mm Browning Long cartridge, a unique cartridge not commonly found today.

Many of these pistols have been converted to fire .380 cartridges. This weapon was adopted by the Swedish army in 1907 and used through World War II. In addition, the weapon was used by the pre-World War II Belgian and Turkish armies, among others. It was also found in the European prewar police departments. The M1907 was commonly imported into the United States in the 1950s.

The Husqvarna-manufactured specimen I tested was in mint condition, despite being at least 60 years old. This weapon has an internal hammer and a grip safety, which though commonly

SPECIFICATIONS

Name: M1907 Swedish

Caliber: 9mm Browning Long

Weight: 2 lbs.

Length: 8 in.

Feed: Single-column box magazine

Operation: Blowback

Sights: Blade/notch

Muzzle velocity: NA

Manufacturer: Husqvarna

Status: Obsolete

thought to be pretty useless is actually quite helpful on this weapon. You do not have to worry about it going off even if the side safety is disengaged because it can't fire until you actually depress the grip safety. The safety is quite flat, however, and difficult to disengage quickly. You must push in with your thumb to make certain that you get sufficient purchase to push it out of the way. If the weapon was yours, you would probably want to build the safety up, but it is no worse than that on a factory-equipped Browning P-35 until just recently. For proper use, both must be built up.

The magazine has a butt release, which is not nearly as efficient as one would like. But unlike on many European automatics, when you pull the Browning's magazine out, it does not cause the slide to close, thereby increasing your reloading speed considerably. The weapon is flat and free of projections, which allows it to be carried easily.

The designer of this pistol, John Browning, obviously understood what the combat pistol was for, unlike so many pistols today that are designed basically to meet target shooters' demands and have sharp edges and ridges all over them that slow down your draw. His design features a rounded rear sight, and the whole weapon is free of all sharp edges and projections.

The Model 07 has a single-column magazine, chambered for ammunition less powerful than the 9mm Parabellum cartridge. Today, it would not be considered a powerful enough load for military purposes. The light recoil permits fast repeat shots. The grip is slightly smaller than the one found on the Colt Government Model and does leave my hand slightly cramped. The front sight is narrow and short and does affect indexing on the cinema combat range. Nevertheless, it did catch the light quite well even in dim light. Although added to aid formal target work, the slight undercutting on the front sight has the effect of throwing all available light on the sight, which helps indexing on the combat range.

Group sizes with this particular weapon ran just under 5 inches, which is twice the group size fired with the M19 on the average, but the weapon did perform quite well on the cinema range.

The M1907 is not the most efficient military pistol, but it is still quite good, if you can accept

The author shooting the M1907 Swedish on the formal range.

The lever on the safety system on the M1907 is conveniently situated, but it needs to be larger.

the caliber limitations, internal hammer, and the safety. I would certainly prefer it to any Luger pistol ever made.

A rear view of the M1907 reveals the rounded rear sight, which is especially useful for a combat arm.

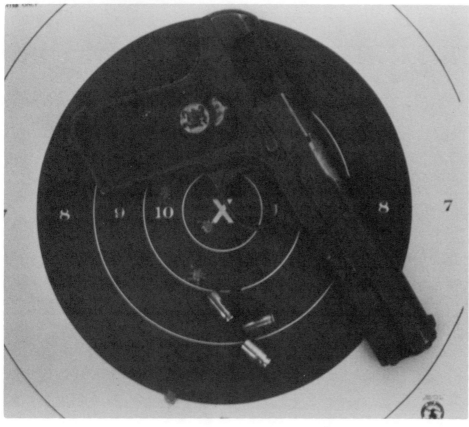

The Swedish M1907 with its 50-foot test target.

M/40 Lahti

9x19mm
(9mm Para)

This 9mm pistol is a standard service weapon in both Sweden and Finland today. It was designed to meet the demands of the Finnish climate and has an accelerator in it that allows the bolt to work even under low-temperature conditions, which can sometimes jam lesser weapons.

This weapon is quite heavy and feels much like a Luger pistol, except that it is butt-release only and the safety is much easier to disengage. Whether by accident or design, the safety on this pistol is very well designed for rapid removal and is much quicker than that of a Luger. By pushing the safety between the thumb, first knuckle, and web of your hand, you can disengage and reengage it quite rapidly, and you can also reengage it with your hand on the weapon in a firing position.

The magazine in the example tested was hard to load. It really needs a loading tool; after three

SPECIFICATIONS

Name: M/40 Lahti

Caliber: 9mm Parabellum

Weight: 2.4 lbs.

Length: 10.7 in.

Feed: 8-rd., in-line, detachable box mag.

Operation: Recoil; semiauto

Sights: Front barleycorn; rear U notch

Muzzle velocity: 1,250 fps

Manufacturer: Husqvarna Vapenfabrik A.B.

Status: In service

or four rounds, it is almost impossible to push the follower down without pushing it against something. In winter conditions with mittens on, I think it would be almost impossible to load the magazine for this particular pistol.

The high front sight mounted on a small-diameter barrel does allow easy indexing, but the rear sight of the pistol is too small and too low for proper use. Also, the back of this pistol is cluttered with a lot of mechanisms, which I found quite distracting when I fired it. Painting the rear and front sights white would make it much faster and less-cluttered looking. Of course, in an Arctic environment, a white front sight might be counterproductive.

The example that I tested was in brand-new condition. The formal shooting, however, was quite disappointing with or without a stock because groups were not very good. I don't know whether this is inherent with the pistol or is due to the ammunition I was using (Winchester Western full-metal-jacket loads). It is conceivable that the barrel bore was slightly bigger than the bullets and was the reason for the poor groups, although no keyholing was apparent. Thinking that it was my shooting the first day, I duplicated the test a number of times with the same results. Group sizes ran about 3 3/4 inches, compared to the average group sizes of 2 1/2 inches I fired with Model 19. Sizes were the same with or without the stock. The weapon had a shoulder stock lug, and I fitted it with a stock. When you use the two-handed Weaver stance, using a stock has few advantages.

The rear sight in this particular pistol is U-shaped, and this makes it hard to hold proper windage in formal target work. It is also quite shallow. Because of the heavy weight of the weapon, recoil is light, and it feels solid in my hand.

Although not the best 9mm by any means, the Lahti is certainly adequate. It does have a few shortcomings. Its accuracy is suspect, but I have heard reports of other examples doing much better. Because it has an enclosed-hammer design, one never quite knows if it is loaded or not. Magazines are hard to load. On the other hand, indexing ability is great, and safety engagement and disengagement is much better than on a Luger pistol.

Of course, this weapon is much too intricately machined for modern-day armies, but if you have a Lahti pistol, you are not bad off.

The author shooting the stocked M/40. Note the fired case exiting the action.

Left side of the Lahti M/40 9x19mm.

A close-up of the accelerator on the M/40. When the temperature drops and the powder burns more slowly, the accelerator helps to compensate, thus ensuring functioning. The accelerator is designed to allow the weapon to operate in subzero temperatures.

A rear view of the M/40 gives a cluttered appearance. This slows down the shooter's ability to pick up the rear sight (above).

Disengaging the safety on the M/40 is difficult if done with any speed (above right).

Completion of the safety disengagement (right).

The stockless M/40 Lahti with 50-foot test target.

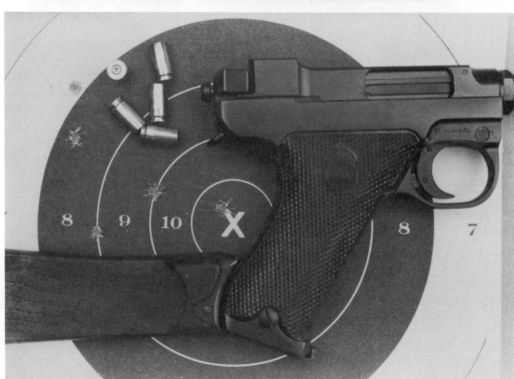

Test target results at 50 feet with a stocked M40.

Smith & Wesson M12 Military and Police, 2-inch "Airweight"

9x29mmR
(.38 Special)

In going through a German-language reference book that deals with military revolvers, I recently found to my surprise that the Swedish air force had adopted and continues to use the Smith & Wesson Military and Police 2-inch-barrel revolver as its standard revolver. Perhaps it did so after the U.S. Air Force lightweight short-barrel revolvers trials in the 1950s. (As a result of those trials, both the Colt "Air Crewman" and the Smith & Wesson M13 were ultimately adopted by the weight-conscious U.S. Air Force. The Air Crewman and M13 were basically alloy versions of the Colt Detective Special and Smith & Wesson Military and Police. Unlike their later commercially available counterparts—or the Swedish duty weapon—these two had magnesium cylinders that made them even lighter, but apparently they were unable to withstand sustained fire and were withdrawn and destroyed.)

SPECIFICATIONS

Name: S&W M12 M&P Airweight

Caliber: .38 special

Weight: 1.1 lbs.

Length: 8 1/4 in.

Feed: Revolver

Operation: DA

Sights: Blade/notch

Muzzle velocity: NA

Manufacturer: Smith & Wesson

Status: In service

Test results for this weapon were similar to those for the steel-framed variant except, of course, recoil was greater. Naturally, no .38 Special revolver, especially with the typical M41 military load of 130-grain, FMJ bullet at 900 fps will be too heavy, but there was a distinct difference between the recoil of these two.

Capacity on all of these Military and Police revolvers is six, one greater than on the smaller J-framed models, but in return you get a bigger and wider pistol. A better trade-off, however, is the action; the K-frames all have the flat mainspring rather than the coil spring of the J-frame. The trigger action is typically smoother and easier to control.

The sights on the M12 are fixed, but the rear sight is nice and square-cut and deep enough to allow a good sight picture. The front sight is ramped and wide to permit rapid indexing in poor light.

On the cinema range, both this weapon and the

steel-framed version proved themselves to be fast and accurate. As one who has been involved with handguns since the early 1960s and who has a background in police work, I never had much use for a 2-inch-barrel K-frame. I thought 2-inch J-frames were quite nice, and I really liked 4-inch K-frames, but a 2-inch K-frame always seemed bigger than you want for a pocket pistol but too short to be easily shot. I realize that people the 1920s, '30s, and '40s did not have any J-frames to work with, and they used the 2-inch K-frame. After working with both the alloy- and steel-frame versions, I have concluded that they make good military weapons for certain purposes (e.g., for an aircrew member who needs something light, fairly small, and easy to shoot that doesn't require much formal training or upkeep). They are much better than any .32 CP or .380 pocket pistols commonly encountered in such roles. So instead of turning up my nose at the Swedish air force's selection, I can recommend this handgun as a fairly intelligent choice.

The author shooting the Smith & Wesson M12 2-Inch .38 Special revolver.

The left side of the Smith & Wesson M12.

The Smith & Wesson M12 2-Inch is the issue weapon for the Swedish air force. The results of the author's 50-feet target test.

M1882 Ordnance Revolver

Switzerland was the first major country to adopt the self-loading pistol for its army, but it also used the revolver for quite a long time after adopting the M1900 Luger. Although the Swiss had adopted earlier revolvers and updated the M1882 in 1929 for simplified production, the point can be made that the M1882 was the handgun actually used the longest by the Swiss military.

The M1882 is an all-steel six-shot weapon chambered for the 7.5 Swiss round, the same round later used in the Swedish and Norwegian 7.5 revolvers, although these differ in design from the M1882.

Bear in mind when we are discussing this weapon that it was designed only nine years after the Colt M1873 .45 revolver; it is not a contemporary of a Smith & Wesson M66 Combat Magnum. The M1882 was originally intended to be used by officers only, but by 1912 Swiss regulations had

SPECIFICATIONS

Name: M1882 Ordnance Revolver

Caliber: 7.5mm Swiss

Weight: NA

Length: NA

Feed: 6-rounds

Operation: DA

Sights: Front bead; rear U notch

Muzzle velocity: NA

Manufacturer: Waffen Bern and SIG

Status: Obsolete

changed so that officers carried the Luger pistol. The M1882 was used by sergeants and corporals of the cavalry, transport, and train units, and by the bicycle troops. Obviously, the objective was to have a weapon that could be used easily with one hand and one that would be free from accidental discharges. To accomplish this, the designers gave the weapon a very sturdy, heavy main spring. Single-action pull on the tested weapon was about 8 pounds, and the double-action pull went 26 pounds. The pistol was fitted properly with the parts numbered, so the heavy pull was not a matter of ill-fitting parts. In tests using the weak hand, a shooter could only snap the weapon 31 times in 30 seconds, whereas he could snap an M65 Smith & Wesson 63 times in the same amount of time.

Besides the extremely heavy pull, which limits the ability to fire quickly on the cinema range and makes formal target work difficult, the weapon's light caliber is also a real handicap. Of course, at the time people did not know that such small calibers wouldn't be successful in handguns since they had worked well in rifles, but today we know better.

The sights on the weapon also leave something to be desired. The front sight is a bead, which makes it difficult to hold elevation on the target properly. The rear sight is a small U notch, leaving very little room for light around the bead. This is no doubt helpful for long-range target work, but, given the trigger pull, this weapon is not really designed for such work.

The pistol does have two excellent features: a swinging side plate and its extraction system. By turning the right-hand frame-screw out, the whole action of the pistol can be opened. This makes parts replacement and cleaning easy and thorough. As anyone who has ever tried to remove or replace the side plate on a Smith & Wesson or Colt will attest, this swinging side plate system whether found on a Swiss or Japanese Type 26 revolver is wonderful. The easier a pistol is to maintain, the more likely it is to be properly maintained.

Although the M1882 is a solid-frame revolver, the cylinder does not swing out. Instead, the ejection rod, which is not spring loaded, must be pushed to the rear to eject the cartridges. However, unlike on a Colt SA Army, the shooter does not have to turn the cylinder by hand. Instead, all he needs to do is pull the cylinder latch down. This disconnects the hammer from the sear so it does not retract when the trigger is pulled, yet the cylinder will turn when the trigger is pulled. The shooter must then simply pull the trigger and punch the ejection rod to the rear (which kicks out the empty), pull the rod to the front until it stops, pull the trigger, and then repeat the whole process. This makes ejection very fast. In fact,

when I discovered this feature, I ran a couple of cylinders of ammunition through the weapon and tested the unloading feature. After a few tries, I was able to unload it completely in a little less than three seconds, which is perhaps not as fast as with a Webley Mark IV but is a whole lot faster than with a Colt SA Army pistol.

On the cinema range, the ropelike grip that appears too small actually felt quite good and allowed the weapon to be easily handled. The bead front sight indexed well on the darkened range, but the heavy trigger pull still presented problems.

The Swiss seemed to like the weapon; they kept it in service until 1964 when stocks were declared surplus, giving it an active service life of 82 years! This is certainly a credit to the design, although, admittedly, the weapons were never really battle tested. All the same, I would suppose two generations of Swiss soldiers would have discovered any flaws in the weapon by 1929 when the weapon was updated for ease of production in the Model 1929 (82/29). Nothing much seems changed.

Swiss handguns are always rare, and you are not likely to encounter one in some distant land, but who knows? All Swiss males are part of the militia, and Swiss citizens do get around. However, you are most likely to find this weapon in Switzerland, where gun laws are more liberal and old revolvers from the nineteenth century are not likely to attract much attention.

The terrible trigger pull is the M1882's biggest fault. Except for that flaw, I would rate this pistol about the same as the Lebel M92. If I could get the trigger pull reduced somehow, I would prefer a Swiss M1882 to a 7.62mm Nagant—and the Russians fought two world wars and a civil war with the latter. There may still be work for an M1882.

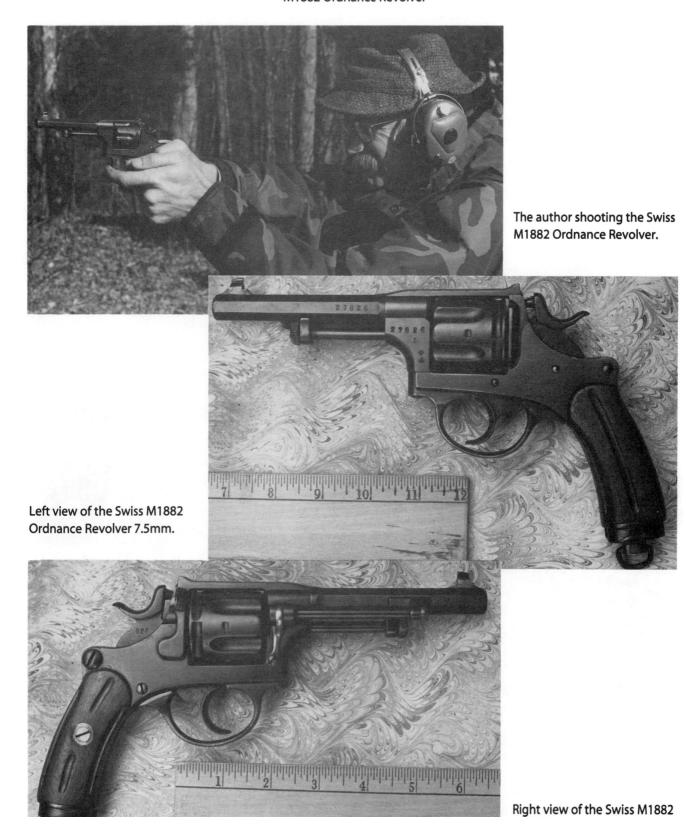

The author shooting the Swiss M1882 Ordnance Revolver.

Left view of the Swiss M1882 Ordnance Revolver 7.5mm.

Right view of the Swiss M1882 Ordnance Revolver 7.5mm.

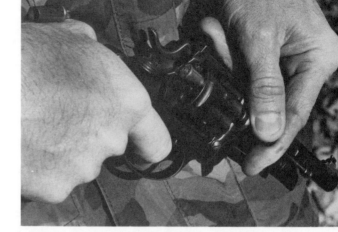

To unload an M1882, simply pull the trigger to rotate the cylinder push rod to the rear. The author has unloaded the weapon in just three seconds.

The open action on the Swiss M1882 is an excellent feature.

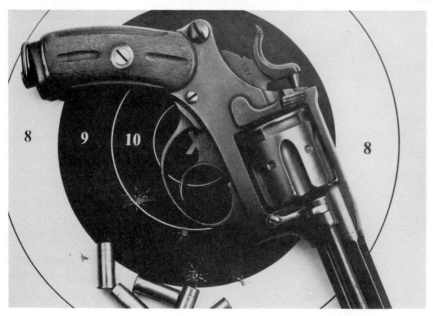

The 50-foot test target with the M1882 Ordnance Revolver.

M1906/29 Luger

7.65x22mm
(.30 Luger)

The Swiss were the first to adopt the Luger pistol in 1901, even before the German navy adopted it. Over the years, the Swiss bought the guns from Germany, then bought licenses to make them in Switzerland. In 1929, they decided to modernize the design to make it easier and cheaper. Looking at the 1906/29 model, I suppose it was a lot easier to make than the 1906 model, but a Colt Government Model in 7.65mm would have been easier and better.

However, the Swiss had two things going for them. First, they are well educated, excellent workmen, and used to the idea of maintaining things; and since the militia members were fighting for their own neighborhoods, they had a real incentive to take care of the weapons. Second, they did not have to prepare to fight all over the world, only in Switzerland, so they did not have to worry about desert or jungle conditions.

SPECIFICATIONS

Name: M1906/29 Luger

Caliber: 7.65mm (.30 Luger)

Weight: 2 lbs.

Length: 9.5 in.

Feed: 8-rd., in-line, detachable box

Operation: Recoil; semiauto

Sights: Front blade; rear V notch

Muzzle velocity: 1,200 fps

Manufacturer: Luger

Status: Obsolete

Despite these advantages, the Swiss-designed Luger suffers from the same problems found in its German cousin. The safety is hard to disengage and impossible to reengage without breaking your grip. One could carry it without the safety on and rely only on the grip safety, but I am always hesitant to do that with any striker-fired weapon and I cannot believe many Swiss officers felt differently.

The sights are the typical barleycorn front and inverted V rear, always a bad combination at least. But the 1 1/2-inch group produced at 50 feet shows that if the shooter does his part, the weapon will perform. All the Swiss guns tested for this volume were excellent weapons.

By the time this weapon was produced, technology had rendered the Luger design obsolete. There were much better designs available both from the standpoint of manufacturing and fighting. One can only conclude that the Swiss stuck by the design because they had a factory already set up to make them in one fashion or another, had a lot of them in inventory, were familiar with them, and liked them. I like looking at them, too, but I certainly would not carry one. The same thing goes for the 7.65mm Luger round. The caliber requires a locking system due to pressure, yet even when this pistol was designed people knew the caliber was deficient in stopping power: the Germans had abandoned the round in 1908 in favor of the 9x19mm because of these concerns. It would take the Swiss another 20 years to come to this

The author shooting the Swiss M1909/29 Luger.

conclusion and adopt the SIG P210 9x19mm. Throughout World War II, the pistols that stood guard over Switzerland were 7.65mm Lugers in three versions and 7.5mm Swiss revolvers. People who like the Luger all like the Swiss version, so you may encounter one in all sorts of odd places.

Right view of the Swiss M1909/29 Luger.

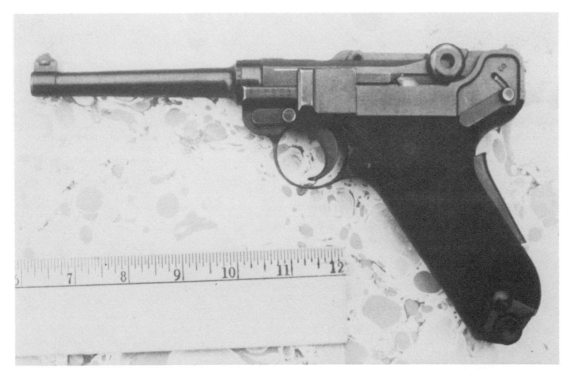

Left view of the Swiss M1909/29 Luger.

The 50-foot test target with the Swiss M1909/29 Luger.

SIG P210

9x19mm
(9mm Para)

Without a doubt, this has got to be the best 9mm pistol in the world. After World War II, the Swiss were looking for a new weapon to replace the 7.65 Luger and the 7.5 Swiss revolver. After much modification, they adopted the SIG P210, also known as the SP 47/8, which was developed from the French Model 1935 A.

The weapon is probably too well made for average military purposes, and even the Swiss no longer procure it for the military. Instead they have the P220, which they call the P75. The SIG P210 has also been used by German border police, the Danish army, and minor league countries around the world, and it is stocked throughout the world by sporting goods dealers. As a target pistol, it is sometimes more acceptable (and accessible) to civilians than a standard military weapon. It is also available as a .22 conversion unit, which can be used in 7.65

SPECIFICATIONS

Name: SIG P210

Caliber: 9mm Parabellum

Weight: 2.1 lbs.

Length: 8.5 in.

Feed: 8-rd., in-line, detachable box mag.

Operation: Recoil, semiauto

Sights: Front blade; rear V notch

Muzzle velocity: 1,150 fps

Manufacturer: Schweizerische Industrie-Gesellschaft

Status: Current production

Luger and is common in Italy as a consequence.

This is an all-steel 9mm pistol, with single-column magazine and single-action trigger. The safety falls readily to your thumb, although the grips for my thumb were a little too thick to allow me to disengage the safety perfectly. By grinding down the grips, as I did on my personal P210, you can disengage it without building up any platforms, which add bulk. If your thumb is a little bigger, you probably would not have that problem; you could also build up the safety if you choose. Grinding the grip material down is just as effective and avoids any extra cost or consequences.

The SIG P210 has one of the best trigger pulls in the world. Groups fired with the SIG P210 at 50 feet were as small as 1 3/8 inches. When you consider that I typically fire 2 1/2-inch groups with the Model 19, it says a lot for the SIG P210. It is such a good pistol that I find it can substitute for a short rifle in my hands. With my personal P210, I have put approximately 18,000 rounds through it at this point in time (most of it Czechoslovakian steel-jacketed surplus or Spanish surplus ammunition). I am able to hit chest-size targets at 300 yards off-hand and do it four times out of eight. At 200 yards I can generally hit it every time.

If you are a well-trained person, you are probably better off with the SIG P210 and good ammunition than the average terrorist armed with an AK47 firing full auto at the same distances.

The safety is somewhat hard to disengage when the pistols are brand new. My personal pistol has had thousands of rounds through it and the safety flipped off thousands of times, so it works properly without any problems. I also tested my friend Leroy Thompson's P210: his was almost new, and the safety was very stiff. However, removing the safety and buffing it just a little on hard stone remedies this problem.

This pistol does have a butt magazine release, which is slow to use and easy to disengage in the car seat when you carry the weapon under your shirt. It happened to me once. Other than that, there is little else wrong with this pistol. Sights are black and should be painted white for fast pick up on the cinema range, but that can be said for almost every weapon tested.

The magazine safety allows you to pull the magazine out and render the weapon safe. It can be cocked and loaded with the safety on, which is nice for military purposes. Its safety is easy to disengage; its trigger pull is superb, which makes

The author shooting the SIG P210 9x19mm.

it easier for people to learn to shoot it well; and it is highly accurate. If you like single-column 9mm pistols with single-action trigger pulls and are willing to pay the price, this is clearly the best 9mm pistol available, particularly if paired with ammunition such as THV, which my SIG P210 fires perfectly.

Left view of the SIG P210
9x19mm.

Right view of the SIG
P210 9x19mm.

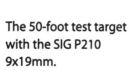

The 50-foot test target
with the SIG P210
9x19mm.

SIG-Sauer P220

9x19mm, 9x23mmSR, 11.43x23mm
(9mm Para, .38 Super, .45 ACP)

The SIG-Sauer P220 is the service weapon in both Switzerland and Japan. It has also been adopted and used by some French army units. The P220 was designed to replace the SIG P210 once it became too expensive to manufacture, even for the Swiss army. The standard weapon is in 9mm, but it is also available in .45 ACP and .38 Super and is marketed in 7.65x22mm in Italy.

All SIG pistols and products are fine weapons, but this one is particularly impressive. My groups were so remarkable that I tested again at 25 and at 50 yards—and the results were just as superb. I placed five shots into a little more than 2 inches, and I pulled one of those shots. Four shots went into roughly 1 1/4 inches. When you consider that my groups with the standard Model 19 Smith & Wesson were 2 1/2 inches, you get an idea of how accurate the P220 is.

The one that I tested was in .45 ACP, and in

SPECIFICATIONS

Name: SIG-Sauer P220

Caliber: 9mm Para (.38 Super, .45 ACP)

Weight: 1.9 lbs.

Length: 7.8 in.

Feed: 9-rd., in-line, detachable box mag.

Operation: Recoil; semiauto

Sights: Front blade; rear notch

Muzzle velocity: 1,132 fps

Manufacturer: SIG-Sauer

Status: Current production; in service

that caliber it was reliable even with Hague Peace Conference ammunition. Although it does have a decocking lever, the type it has is the generally less desirable nonselectable, double-action version. However, for military purposes, this probably makes little difference. It lacks a magazine safety, but it is adequate without it.

The smooth finish on the trigger, combined with the gun's light weight and .45 ACP caliber, did cause the trigger to break away when the weak hand was used to fire. Roughening the trigger surface slightly would cure this easily enough.

On the cinema range, this weapon also performed quite well. The wide white front sight and rear sight allow good indexing in poor light. The double-action pull is smooth, and although it is out slightly too far in double action, it can be accommodated easily enough. The P225 is a little better in this regard because of the shorter overall trigger reach. The safety controls fall easily in the hand. The aluminum frame results in a very lightweight weapon.

The heel-mount magazine release on the P220 slows reloading, but reloading speed is not usually a critical consideration with military weapons anyway. Further, the U.S. market now has the P220 available with the side-button magazine release.

The major advantage of this design is its availability in .45 ACP. Despite debate about whether 9mm pistols are adequate for antipersonnel use in police work, in a military situation, Hague Conference ammunition *must* be used. With Hague Conference full-metal-jacket ammunition, the 9mm Parabellum is not an adequate antipersonnel load; in .45 ACP, however, the 230-grain bullet has been shown to provide adequate defense. This is assuming, of course, that you are restricted to regular conventional ball ammunition sanctioned by the Hague Conference rules, and neither expanding ammunition or the new hypervelocity THV round is an option and that ballistic vest penetration is not a serious consideration.

The author shooting the SIG-Sauer P220 .45 ACP. The weapon's light weight (26 ounces) and ball ammunition cause some muzzle flip.

Although the P220 is not as good as a P225, I would rate it as the top weapon that I tested *in .45 ACP.* Unfortunately, to my knowledge, no country has adopted it in .45 ACP; all use it in 9mm Parabellum.

In summary, the P220 is safe, accurate, quick to get into action, has good stopping power and sights, and allows rapid indexing in poor light. I highly recommend it.

Right view of the SIG-Sauer P220 .45 ACP.

Left view of SIG-Sauer P220 .45 ACP.

Front view of SIG-Sauer P220 (right, .45 ACP) and P226 (left, 9x19mm).

The 50-foot test target with the SIG-Sauer P220 .45 ACP.

Nagant Model 95

7.62x39mmR
(7.62 Nagant Revolver)

The Model 95 Nagant revolver was a standard service weapon of the Czarist Russian army during World War I and continued to be fielded by the Soviet army throughout World War II. The example that I tested was a Finnish army-captured weapon manufactured in 1937.

This weapon has one good feature: the high front sight that permits rapid indexing on a combat range. Otherwise, this is a terrible weapon, especially when compared to other revolvers from a similar period. In comparison to the Webley Marks III, IV, and VI, the Nagant Model 95 is clearly inferior.

Reloading is slow; cases are fired by a gas seal system, and they expand. Ejection is likewise slow and a one-at-a-time procedure. It does hold seven rounds, but the load of 7.62 Nagant medium-velocity with a full-metal jacket is not a guaranteed stopper by any means. Recoil is light, but the Fiocchi ammunition produced a lot of flash on the cin-

SPECIFICATIONS

Name: Nagant Model 95

Caliber: 7.62mm

Weight: 1.7 lbs.

Length: 9.06 in.

Feed: 7-round cylinder

Operation: DA revolver

Sights: Front blade; rear U notch

Muzzle velocity: 892 fps

Manufacturer: State arsenals

Status: Obsolete

ema range. On the example I tested, the trigger pull was heavy on single action and double action; the sights were small, dark, and hard to see on formal target work, and the shallowness of the rear sight made it hard to maintain your windage and elevation.

The gas-seal system on this Nagant Model 95 creates a lot of complications with very little benefit. It requires, upon the final cocking of the action, that the cylinder be pushed forward by what would be the bushing in a normal Smith & Wesson revolver. This forces the cylinder over the barrel, thus removing the usual gap there. Theoretically, this design element increases the performance of a given load. Test results in a recent *American Rifleman* show that this system gets fewer than 100 fps additional velocity and at great cost in terms of mechanical complication. Further, because the cylinder has to be pushed forward under pressure from the bushing, this results in a heavy trigger pull.

You could accomplish the same result without this complication by loading your ammunition a little hotter, getting an extra inch of barrel, or using a better caliber.

On the cinema range, the heavy trigger pull and the poor sights contributed to a terrible performance. Out of 15 targets, I hit only three. I fire on the cinema range quite frequently, and I am experienced with moving targets under low-light situations. On the formal range, the Nagant performed better but by no means great: a little less than 5 1/2 inches for five shots at 50 feet, or twice as large as similar groups from the Model 19. It may feel good in your hand, but it is not accurate.

It is often said that Nagant revolvers are crudely manufactured, but the one I tested, which was manufactured in 1937, was not. There was quite a bit of handwork in it, especially in the inlaid grips.

The Nagant Model 95 revolver would be a great weapon for your enemies to have: only seven shots, slow reloading, low-power ammunition, mechanically complex, and hard to use.

The author testing the Nagant M95.

Right side of the Nagant Model 95 7.62x39mmR.

Left side of the Nagant Model 95 7.62x39mmR.

The extremely long hammer nose of the M95 Nagant is apparent in this photo.

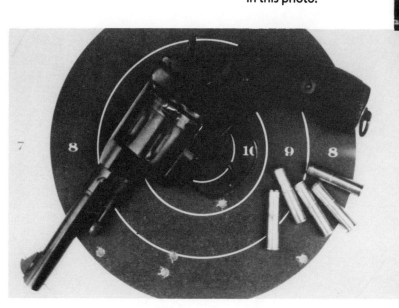

The Nagant M95 50-foot test target.

TT M1933

7.62x25mm
(7.63 Mauser, .30 Mauser)

The Soviet Union adopted this pistol in 1930, made slight modifications in 1933, and retained it as its standard handgun until the 1950s.

The handgun I tested was manufactured in 1966. It was given to me by one of my agents who was formerly assigned to the Phoenix Program in Vietnam and who captured it in the field. It came complete with a holster and spare magazine. Although the finish is worn, it functions without any difficulties.

The grip angle on the pistol (known as the Chi-Com 54) is quite steep. As a consequence, it tends to shoot somewhat low, and you must train your wrist to properly cant it, or you will hit people in the knees instead of the chest. Of course, practice can help you compensate for this.

These pistols do not have any safety on them, so you carry them in condition three, two, or zero. I view this lack of safety as a weakness, but

SPECIFICATIONS	
Name:	TT M1933
Caliber:	7.62mm
Weight:	1.9 lbs.
Length:	7.68 in.
Feed:	8-rd., in-line, detachable box mag.
Operation:	Recoil, semiauto
Sights:	Front blade; rear U notch
Muzzle velocity:	1,378 fps
Manufacturer:	State arsenals
Status:	Obsolete

perhaps it forces people to rely on their brains and to be careful. I rather doubt it, though, and imagine that a lot of accidental discharges took place whenever this pistol was in common use.

The magazine on this pistol is sturdy because the feed lips are in the magazine housing, thus avoiding most of the magazine-related malfunctions related to bent feed lips.

The caliber of 7.62mm, which is virtually identical to .30 Mauser, is an interesting choice. Apparently, the Soviets had a lot of Mauser pistols in that caliber and liked them. Additionally, they could use the same barrel-making machinery for their pistol barrels that they used for their rifles. The .30 Mauser is not known as a stopper, but it does offer good penetration. For Soviet troopers facing winter-clad Germans with their MP40 magazine pouches across their chests, this penetration may have been quite useful. Also, because the Soviets used the same caliber in the PPSh 41 and

PPS 43 submachine guns, they developed a number of interesting nonball rounds. Russian ammunition that traced, had incendiary capability, and was armor piercing was available. I would have preferred incendiary ammunition for antipersonnel use if possible.

Sights on the pistol are quite good. The rear sight is narrow but high, and the rear sight is as good as the custom sights found on many combat match pistols today. This allows rapid indexing on the cinema range but would be even better if painted white. Trigger pull is typical GI (i.e.,

awful), but one really did not notice it on the cinema range. It did make fine work on the formal range difficult.

Wherever these pistols are found (and they can be found almost anywhere), they are known to be heavy duty and reliable. The finish is standard blue, which leaves a lot to be desired, but I guess it was the best the Chinese could do at the time. Watch out for the lack of safety, paint your sights white, try to get some effective ammunition, and you will have a fairly effective military handgun.

Chi-Com Type 54 Tokarev captured in Vietnam and presented to the former head of SOG (Special Operations Group) in 1972.

This TT 33 7.62x25mm is actually a Chinese Type 54 captured in Vietnam.

The 50-foot test target for the TT 33.

Makarov

9x18mm
(9mm Makarov)

The example that I tested was a Type 59 Chinese model manufactured in 1966 and captured in Vietnam. In fact, it was still as full of grease the day I tested it as it was the day it came back from Vietnam. I had to clean out the grease with gasoline to fire it.

The 9mm Makarov ammunition was quite hard to obtain here in the United States at the time of the test, so I tested it with both 9mm Police (Fiocchi) and .380 (factory Remington) ammunition, as well as Makarov ammunition. The fact that the gun functions with .380 (which is commonly available) or 9mm Police (which is even more readily available, at least in Western countries) is something to keep in mind, but accuracy does suffer. With .380 ammunition, I had groups slightly greater than 7 1/2 inches for five shots. The bullet is too small by some .005 of an inch. It also fired about 6 inches high at 50 feet; with proper ammunition, group sizes would

SPECIFICATIONS

Name: Makarov

Caliber: 9mm

Weight: 1.6 lbs.

Length: 6.34 in.

Feed: 8-rd., in-line, detachable box mag.

Operation: Blowback, semiauto

Sights: Front blade; rear square notch

Muzzle velocity: 1,070 fps

Manufacturer: State arsenals

Status: Current production; in service

probably be much smaller.

The front sight is small and black, but it did pick up well because of its square configuration. Both sights need to be painted white to allow fast indexing, although the rear sight did pick up surprisingly well on the cinema range despite its small size. The rear sight should be rounded off to avoid snagging on clothing.

The safety on this particular pistol is the opposite of the Walther. You flip it up to engage and flip it down to disengage. Once the safety is on, the slide cannot be retracted, which I suppose has certain advantages for military weapons. The trigger guard has a squarish shape, which helps with the hold that I use where my weak-hand index finger is wrapped around the trigger guard. The pistol generally feels good in my hand, and, unlike with the Walther PP/PP series, the slide does not bite into the web of my hand because it is higher over my hand than on the Walther. Although bigger, the Makarov feels much like a Sauer Model 38 in your hand.

The Makarov has a heel-mounted magazine release, but, unlike on so many European automatics, the slide does not close when the magazine is withdrawn. The slide stop is easy to disengage when the new magazine is inserted in the pistol. You can hit the slide stop with a thumb and promptly load the weapon. On the example I tested, the double-action pull was heavy, but the single-action was quite light. The trigger broke rather suddenly in double action, and I had to watch it carefully.

The Makarov pistol is chambered for a cartridge of power between the .380 and the 9mm Parabellum. As a result, it is better than any .380, simply because it is about 100 fps faster, but it is clearly not up to the power levels of 9mm Parabellum cartridge. It is, however, smaller than many 9mm pistols, and that gives it certain advantages. It needs to be chambered for a more potent cartridge, but by using ammunition such as THV or Arcane, you might be able to overcome this problem.

The magazine is cut out on the sides to allow you to clean the mud and crud out, which certainly is desirable on a military pistol. Another plus is the fact that both the bore and the chamber are chromed, especially useful on military weapons subjected to low maintenance and extreme conditions. Oddly enough, however, the weapon does have a high-polish blued finish. A

more durable flat or matte finish seems more practical, particularly since the designers chromed the bore and chamber to lower maintenance require-

The author testing the Makarov 9x18mm pistol.

Makarov 9x18mm pistol with Soviet airborne paraphernalia.

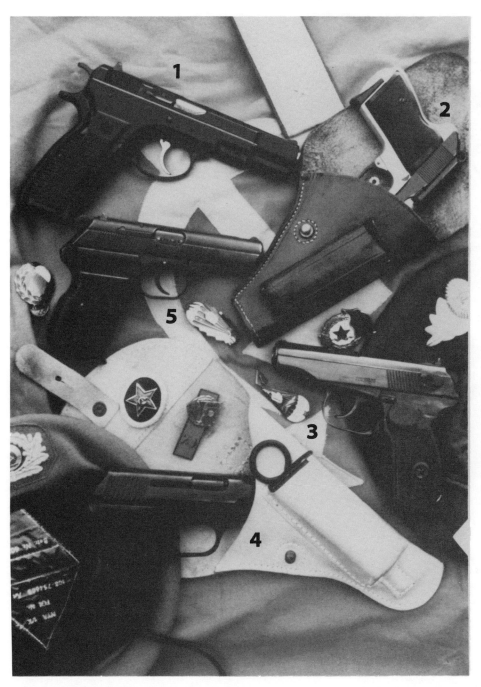

ments. The example I shot apparently had different alloys in the hammer and safety; they had taken on a pink finish in reaction to the blueing salts.

Generally, this is a well-made and well-assembled pistol, not at all crudely finished. Some hand fitting was evident when I disassembled the weapon to clean it, and all parts were serial-numbered to the weapon. It is better than any other Soviet military weapon, other than the PSM 5.45mm pistol.

Compared to the Nagant M95 or the Tokarev TT 1933, the Makarov must have seemed a giant step forward to Russian soldiers. Although not as good as the Glock 17 or the SIG P225, it is clearly the equal of many Western European pistols. The fact that you can use .380 ammunition in it is an advantage for those of us in the West.

Handguns of the former Warsaw Pact nations (top to bottom): 1) CZ 75; 2) RK-59; 3) CZ 50; 4) Makarov; 5) Vz 52

Makarov 9x18mm captured in Vietnam and the 50-foot-test target shot with it.

PSM

5.45x18mm
(5.45 PSM)

This pistol is very rare in the West. With the breakup of the Soviet Union, it might become more common. These pistols were first reported to have been designed for general officers. Yet you never seen one carried by a general officer. Odd! Then you read that they were designed for law enforcement agents. But all the law enforcement agents seem to have something else. Even odder! Then you hear that they were used by the KGB. Well, maybe; but Keith Melton, the noted intelligence service expert, states that his high-ranking KGB source says he never even shot one, much less carried one. Curiouser and curiouser! Who did get them, how were they used, and why were they built?

When you first encounter a PSM, you are struck by the flatness and thinness of the weapon. A Walther TPH is smaller both in barrel length and grip, yet it is half again as wide.

SPECIFICATIONS

Name: PSM

Caliber: 5.45 PSM

Weight: 1.2 lbs.

Length: 6.24 in.

Feed: 8-rd. box mag.

Operation: Fixed-barrel blowback

Sights: Front square; rear notch

Muzzle velocity: 1,024 fps

Manufacturer: State arsenals

Status: Current production; in service

The grips are made out of some metal alloy, which is unusual because obviously they would be cold, something you think would be a Soviet concern. The safety is hard to flick on and off, but since this is a double-action design, you do not need the safety except after firing a round. Then you are faced with flicking it on and dropping the hammer or running around with a cocked and loaded pistol ready to fire. Naturally, you can run with your finger off the trigger, but I do not think this is the best solution myself.

The chamber and bore are chrome-plated, in keeping with current Soviet and Chinese practice.

After fully examining this weapon, you still wonder why it was built. The 5.45mm caliber is a unique choice. Why not a .22 rimfire and full-metal-jacket ammunition, as the U.S. Air Force did with its survival rifles if the Hague Conference prohibitions bother you? Certainly the cost of training ammunition would be less. Only after much effort and not an inconsiderable amount of expense was I able to locate enough ammunition to test this weapon. Although the weapon is lightweight and thin, its recoil and noise are both quite light. It feels about like a standard velocity .22 rimfire Long rifle.

Sights are acceptable in that there are a square front and rear notch with no white dots, merely a square post. A dab of white paint would greatly assist pickup, especially on the cinema range.

The weapon shot well, as the 1 9/16-inch group evidences. Still, a 5.45mm bullet does not strike me as really suitable for military handgun use. There had to be more. The Soviets would not go to all this trouble and expense if they did not think they were getting something for it. Remember, the Russians are the world-famous chess people!

Finally, it dawned on me! This weapon was designed specifically to penetrate ballistic vests. I immediately got out my sample of Level 11A vest material, which will stop standard NATO 9x19mm, .357 Magnum loads, and everything of lesser power. The standard-factory-load 5.45mm round went right through and then penetrated 250 pages of telephone directory. I was right: this weapon was designed to shoot through vest material. Its thinness aids concealment. The pistol

The author testing the Soviet-designed and issued PSM 5.45mm pistol.

gives off a metal signature when tested on the metal detector, but it's less than a similar Western steel weapon. But most important, the Soviet rounds all contain a steel penetrator.

Given the introduction of ballistic armor by the Western armies, this weapon goes from being an odd-calibered little pocket pistol to a serious combat pistol. After all, a weapon is only a bullet projector. Because of this unique capability of the Soviet PSM, this weapon gets into the top selection of military weapons. Though my personal preference is a THV-loaded Glock 17 or SIG P225, if I were anticipating a modern, fully equipped enemy, I would choose the Soviet PSM over ball ammo loaded M1911/A1.

The Soviet PSM 5.45 was designed specially to penetrate ballistic vests worn by NATO armies (left).

The 50-foot test target shot with the PSM 5.45mm pistol (below).

Colt M1909

11.43x32.1mmR
(.45 Long Colt)

After the adoption of the Colt .38-long-caliber revolver turned out to be a disaster, the U.S. military went looking for an alternative weapon. They recalled a lot of Colt SAA pistols in .45 Long Colt caliber, refinished them, and shortened the barrels to 5 1/2 inches, but everybody knew this was only a temporary expedient. What the military wanted, of course, was a modern semiautomatic pistol. However, in the early 1900s, none of the ones available in a large caliber were reliable. The U.S. military had learned its lesson about small-caliber weapons, even if its European counterparts (except in Britain) did not understand. The United States wanted a .45!

The call went out, and many pistol "designers" came forward. John Browning's M1905 was nice, but it did not shoot a 230-grain bullet, lacked some safety features thought desirable, and, of

SPECIFICATIONS

Name: Colt M1909

Caliber: .45 Long Colt

Weight: 39 oz.

Length: 10 3/4 in.

Feed: Revolver

Operation: Double action

Sights: NA

Muzzle velocity: N/A

Manufacturer: Colt

Status: Obsolete

course, really was not as reliable as necessary. While John Browning continued to work on (and ultimately prevailed with) his M1911, the U.S. military needed a modern large-bore handgun.

Colt New Service revolvers had first been made in 1898. Soon they became the obvious choice for any U.S. citizen who sought a large-bore weapon but wanted something more advanced in design than the Colt SAA. At the time, Colts also led the market in personal defense weapons.

Although I would have preferred the products of Smith & Wesson, that company did not make a large-frame, large-caliber, solid-frame revolver until 1908. During the period from 1898 to 1908, the Colt New Service was king, and the king continued to have many followers until it went out of production in 1943.

The M1909 is basically a standard Colt New Service 5 1/2-inch-barrel handgun chambered for .45 Colt with some military-specified grips and finish. It has all the good and bad points of any New Service. I find the grip too large for my hands. Also, without some type of adaptor, it shifts in my hand when I fire rapid double-action strings, slowing up overall response time. The sights are high enough on the long cylindrical barrel to make rapid indexing possible, although the low narrow notch in the frame that is the rear sight makes triangulating difficult on the cinema range. A dab of white paint on both would improve accuracy. The single-action pull is good, but, because of the grip size, I find the double-action pull to be heavy.

A more serious problem involves the ejector rod. It hangs out under the barrel unsupported. I have heard reports of the weapon being jammed because of twigs and debris preventing the cylinder from opening. The encased ejector rod of a Smith & Wesson avoids this problem, but it is both more expensive to make and can be clogged with mud, preventing the cylinder from closing. On the Colt, such mud can be quickly cleared, whereas on the Smith & Wesson, it can't. An even worse problem with the ejector rod on the Colt is that of the ejector rod bending when striking a head. During World War I, this apparently was such a problem that manuals had to be published instructing soldiers to hit the enemy with the side of the weapon, not the barrel. Perhaps hitting enemy troops who wore steel helmets, which would tend to bend the rod more than simple bone, were the cause of this concern, but it is still deserves mention. Bashing people with my revolver has never been my style, but it may be yours. If so, watch out.

The M1909 was short-lived in U.S. military circles; the caliber was quickly withdrawn once the .45 CP was developed. These pistols saw use in World War I, but by World War II they were uncommon. Yet, in 1976, when I was ordering ammunition for the federal police agency I headed at the time, I remember reviewing ammunition availability from government arsenal lists and finding .45 Colt ammunition listed. Whether the arsenals really had it or not is another question; perhaps they simply had failed to delete it in the catalog, but I thought it interesting.

I find all these Colt large-frame revolvers too big and heavy for my use. But there is no doubt that they have done everything that you could expect of a revolver, have done it all over the world, and stand ready to duplicate these earlier performances if you find yourself in need of such a weapon.

It was incidents like this that caused the U.S. Army to abandon the .38 and go the the .45 round. Even with close, clear shots to the body, no penetration is depicted.

This detail from a Frederick Remington painting shows an Indian Wars-era scout with pistol.

U.S. troops fighting against Moro guerrillas in the Philippines near the turn of the century. Photo courtesy of the National Archives.

Colt
M1911/M1911A1
11.43x23mm
(.45 ACP)

Since the famous Colt Government Model was adopted by the United States Government in 1911, it has been manufactured in the United States, Norway, Argentina, Canada, and, with modifications, in Mexico, Spain, and elsewhere.

Probably the best thing about this design is its caliber. The .45 ACP caliber is a proven manstopper. The general adoption of ballistic vests today has certainly affected the stopping power of cartridges, but when considering nonvested subjects, the .45 ACP caliber is probably the best handgun antipersonnel round of the bullets that comply with the Hague Conference rules. Certainly, the ammunition is heavy and may produce more recoil than other calibers do, but from the standpoint of stopping an individual quickly with a hit that may not be perfectly placed, it is superior to any other military caliber in existence today—if ballistic vest penetration is not a consideration.

SPECIFICATIONS

Name: Colt M1911/M1911A1

Caliber: .45 ACP

Weight: 2.4 lbs.

Length: 8.62 in.

Feed: 7-rd., in-line, detachable box mag.

Operation: Recoil, semiauto

Sights: NA

Muzzle velocity: 830 fps

Manufacturer: Colt (others in wartime)

Status: Current production; in service

The Government Model is well known to most, but a few observations may be in order. The quality on the Model 1911 example that I tested, which was manufactured in 1915, was obviously higher than that found on the Model 1911A1 version, which was manufactured in 1944. Both finish and the general feel are better. This is not to imply that the 1944 version of the Model 1911 A1 is a piece of junk; it is also well made.

The sights on the 1911, as well as those of 1911A1, are really too small for rapid pickup and certainly need to be painted white. Walter Winans, the famed American target shooter of the early 1900s, who was based in England, noted that he also felt that the Colt Government Model's front sights should be light-colored, preferably white. With his personal 1911, Winans replaced the small, dark front sights with a white front sight to allow better indexing. But he could have resolved the problem with a little dab of white paint.

The weapons are quite heavy at 39 ounces unloaded, and that is a real drawback today. It offers a single-column magazine, but it does fire the .45 ACP cartridge.

Probably the worst thing about this particular pistol for military purposes is that for it to be ready for action, it must be carried cocked and locked. This does present a lot of problems because there are many people who do not know how to carry cocked-and-locked pistols properly. It is very easy for the expert pistol handler to say anyone ought to be able to handle it, but as anyone familiar with military organizations knows, many people are not interested in handguns at all, even though they may be carrying them to save their lives. During World War I, the manuals for the military recommended carrying the Colt Government Model cocked, side safety on. After the war, the manuals recommended that the weapons be carried with the hammer down, chamber empty, and that they by cocked and loaded only when ready to fire. This change could have reflected the skill level of the troops or the number of accidental discharges they experienced. I know the FBI got rid of the Government Models in the 1930s because of this problem, and the FBI agents are better trained than soldiers.

The other problem with this pistol is the lack of a magazine disconnector. People cock the weapon, failing to realize that there is a round in the chamber; then they take the magazine out, and the weapon discharges when they drop the hammer by pulling the trigger. That is why most U.S. military police organizations have 55-gallon drums filled with sand at the door so that people can clear their weapons by firing them into the drum before they go off duty. This is a sad reflection on the skill level found in most military police units.

With eight shots, Sgt. Alvin York managed to stop eight bayonet-wielding Germans who attacked him at 20 yards during World War I. It seems unlikely to me that there was any other pistol at the time (or now—except perhaps a 10mm) with which Sergeant York could have performed this feat. As this illustrates, the caliber is good.

For people who are expert gun handlers and who do not need to carry a rifle, this is probably a fine military weapon. It certainly is one that you are likely to encounter in your travels around the world. They have been widely distributed, thanks to the United States Lend-Lease and Foreign Military Sales programs during World Wars I and II and thereafter. I remember reading a few years ago about someone who encountered a Model 1911 in France that apparently had been left behind in a dugout from 1917. And 60 years later, it was still cocked, loaded, and ready to go. The finish was rusted, of course, from exposure to the elements, but it was ready to fire.

My familiarity with the nature of the personnel who may be assigned a handgun precludes me from recommending Models 1911 and 1911A1 as military weapons because of safety concerns. But they still have a lot going for them. I would not hesitate to carry one myself. The SIG P220 in .45 ACP offers you all the advantages of this caliber but is lighter, safer to carry, and overall a better choice for most military organizations today. There are handguns that are safer, lighter, and easier to carry, but once the shooting starts, there is none better than the Government Models.

It has been said that the .45 CP cartridge was made obsolete by the general adoption of 9x19mm. I disagree. The only logical reason to pick a 9x19mm handgun is because the ammunition for your submachine and handgun interchange. With the general adoption of the assault rifle, the interchangeability of ammunition between your submachine gun and your pistol should no longer be a critical issue. Submachine guns gradually have been taken out of the military arsenal. For police purposes, interchangeability between submachine gun and handgun ammunition seems irrelevant. It is unlikely that any police organization is going to engage in such active, long-term actions that they have to worry about resupplied ammunition. Similarly, the weight savings means little, given the small amount of ammunition called for. The problem of penetrating an armored vest still remains, but rounds such as the THV are available in .45 ACP and could be used to overcome this. This would give the soldier of the future the best of both worlds.

As Jeff Cooper once said, "People shoot 9mms for the pistols and .45s for the caliber." The caliber is still the best thing about the Government Model (and I like Government Models).

During the Mexican Campaign, this U.S. soldier kept his newly issued M1911 close at hand and ready for action by folding the flap back on his holster. Photo courtesy of the National Archives (left).

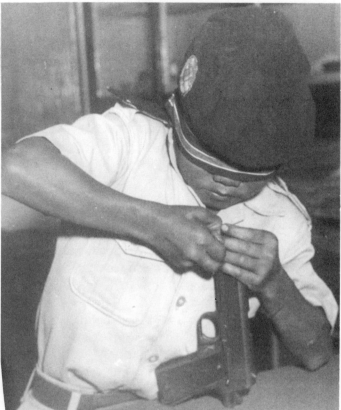

A Vietnamese cadet practices disassembling an M1911A1 blindfolded (above right).

The 50-foot test target shown with the M1911 .45 ACP (right).

The author test-firing the M1911A1, which was one of his favorite handguns, but not a good combat selection these days (above).

Right side of the M1911A1 .45 ACP (above right).

Left side of the M1911A1 .45 ACP (right).

The 50-foot test target shot with the M1911A1 .45 ACP.

Colt M1917

11.43x23mm
(.45 ACP)

When they went into World War I, the U.S. armed forces were ill prepared. They had adopted the M1911 pistol, but supplies and production were limited. General Pershing's announced goal was to have a handgun for every military man in France. To do this while still maintaining the production level of rifles at Springfield Armory and machine guns at Colt meant that some handgun other than the M1911 had to be used.

Contracts were let with other manufacturers to make the M1911, but weapons were needed immediately. Both Smith & Wesson and Colt had large-frame revolvers in production, but none would shoot the rimless .45 ACP cartridge. Faced with a crisis but knowing that the military would not want to introduce another cartridge into the supply lines, Smith & Wesson came up with the half-moon clip idea. By using this, the cartridges could be chambered

---

SPECIFICATIONS

Name: Colt M1917

Caliber: .45 ACP

Weight: 2.5 lbs.

Length: 10.8 in.

Feed: 6-rd. revolving cylinder

Operation: DA revolver

Sights: Front blade; rear square notch

Muzzle velocity: 830 fps

Manufacturer: Colt

Status: Current production; in service

---

in the revolver, and they could be loaded faster than with a normal revolver. The only drawbacks were that the half-moon clips were somewhat flexible, and they tended to cushion the hammer blow. As a consequence, the hammer had to fall harder on a .45 ACP chambered revolver than on a .45 Colt or other rimmed caliber revolver, and the trigger pull was generally harder.

My comments on the Colt New Service style of revolver (page 31) apply to the M1917, which is merely a Colt New Service with a 5 1/2-inch barrel, a rough finish, smooth wooden grips, and a trigger pull that is generally harder because of the half-moon clip.

While no doubt a rugged, heavy-duty revolver, the Colt M1917 is not as good as an M1911, unless you are left-handed. If you are, then either Colt or Smith & Wesson M1917 may work better.

After the war, many of these revolvers were

used by cavalry and military police units. Additionally, the U.S. Postal Service used them because at one time revolvers were common in each postal substation to guard valuable packages. Some M1917 pistols were used as substitute standard handguns or sought out by those who did not like the semiauto handguns during World War II.

In the 1960s, tens of thousands were surplused off, often for as little as $29.95. A Smith & Wesson large-frame revolver sold for $140 at that time, so this was an extraordinary bargain. Most were bought up and modified to suit individual owners' ideas. Consequently, the few that remain unaltered but in good condition are worth more than a new N-frame Smith & Wesson today. This is particularly odd, since the new weapon is clearly superior as a shooter.

People who like the Colt New Service revolver tend to ignore its deficiencies. Although it would not be my choice as a combat revolver today, the M1917 was a welcome development at a time when it was more critical to get handguns into the hands of fighting men than it was to develop the most sophisticated combat revolver.

The Colt M1917 revolver was carried by black troops early in World War II (above).

World War I-era doughboys proudly show off their M1917 Colt revolvers (right).

Smith & Wesson M1917

11.43x23mm

(.45 ACP)

This pistol was a disappointment to me. One of my favorite personal weapons is a Smith & Wesson Model 26 Target revolver in .45 ACP. When I picked up the Model 1917, which I borrowed to test, I anticipated that it would be equal to the Target pistol; however, that was not the case.

The Model 1917 has a double-action pull that must have weighed at least 20 pounds. This weight, coupled with the grip that had no adapter, caused the weapon to shift in my hand with each shot and throw the shots right each time.

Accuracy with the pistol is a problem. During rapid firing on the cinema range, the grip config-

SPECIFICATIONS

Name: 1917 Army Model .45

Caliber: .45 ACP

Weight: 2.5 lbs.

Length: 10 3/4 in.

Feed: Revolver

Operation: DA

Sights: NA

Muzzle velocity: NA

Manufacturer: Smith & Wesson

Status: Obsolete

uration caused the weapon to shift in the web of my hand, reducing efficiency considerably. Cases tended to stick in the chamber even though I was using commercial full-metal-jacket and military ball loads. The three-shot half-moon clips were difficult to remove in a hurry. As a result of all these factors, my groups were about 4 5/8 inches, or roughly twice those of the Model 19 Smith & Wesson fired the same day.

All in all, the Model 1917 is a very disappointing weapon. It would appear to be a safe military weapon, but one is hard-pressed to see that it is as good as many other big-bore revolvers. If you have a choice between this and the Webley Mark VI .455, pick the latter.

The author test-firing the M1917 Smith & Wesson revolver (above left).

The left side of the Smith & Wesson M1917 .45 ACP revolver (above).

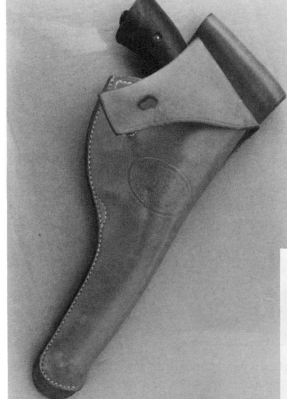

Standard-issue World War I- and World War II-era holster for both the Colt and Smith & Wesson M1917 revolvers (left).

Smith & Wesson M1917 .45 ACP revolver with 50-foot test target (below).

Colt Model 1908

7.62x17mmSR

(7.65 Browning, .32 ACP)

The U.S. Army has issued Colt pocket pistols in .32 ACP or .380 caliber at various times as general officers' weapons. They also have been used as nonstandard weapons for a variety of intelligence agents and undercover officers in the United States and foreign military forces. They were quite common in the hands of British Special Operations Executive (SOE) agents during World War II. This clearly qualifies the pistol as a military weapon.

The example I tested was well worn. Its sights were too small, and the rear sight was much too narrow for proper sighting. Its front sight was also quite shallow in height, which makes it difficult to pick up rapidly in poor light. The trigger pull ran about 15 pounds. The weapon's sights were not properly adjusted, and this made it fire approximately 2 inches high, about 8 to 10 inches left at 50 feet. Still my five-shot group did reveal four shots within 1 7/8 inch,

SPECIFICATIONS

Name: Colt Model 1908

Caliber: 7.65 Browning (.32 ACP)

Weight: 1.5 lbs.

Length: 6 3/4 in.

Feed: Single-column box magazine

Operation: Blowback

Sights: Front blade; rear notch

Muzzle velocity: N/A

Manufacturer: Colt

Status: Obsolete

with the fifth one pulled out to 5 5/8 inches. This pattern seemed to result from something inherent in the weapon: several times I placed four shots within a fairly good group with the fifth out of the circle. I suppose that there was something about the barrel lockup or bushing that caused this problem.

The recoil on this weapon was quite light, despite its heavy trigger pull that made it difficult to handle on a formal target range. On the cinema range, the trigger pull did not affect results, which were quite good. Interestingly, however, when using Western 71-grain full-metal-jacket ammunition, the flash was quite high in comparison to the low power of these loads. This could have resulted from the powder charge of that lot of ammunition because other ammunition tested in .32 ACP did not seem to have as much flash.

Results on the cinema range were fairly good because the grip is very similar to that found on the Government Model or the SIG P210, with which I am quite comfortable. The sights on the cinema range were hard to see and, because they were dark, difficult to properly index.

The safety falls under the thumb for rapid disengagement and reapplication, although it is quite thin. You have to push in using considerable pressure with your thumb to get it off with certainty. Although I do not care much for .32 hammerless automatics, I can see how the slick profile would make it a good weapon undercover agents—though not a very suitable general officers' weapon.

The magazine is difficult to insert properly because of the small size of the hole at the bottom of the grip. This, coupled with the grip of the magazine on the rear of the magazine (which is its release), slowed reloading. A side-button release and a flared housing would speed this up.

It is really hard to see how this is a better weapon than a .45 Colt Commander if a suitable holster is available. It does not have a suitable power level, and it is not as safe as desired for military purposes. Although it certainly has been used over the years by a wide variety of people, it is not a very suitable military weapon.

A Korean general armed with the M1908 Colt greets a U.S. military unit.

A U.S. general officer in Vietnam armed with an issue general officer's Colt Pocket Model M1908 (right).

The Colt M1908 .380 with some British wartime items associated with SOE (far right).

Right side of the Colt M1908.

Left side of the Colt M1908.

The Colt M1908 .380 with 50-foot test target.

Remington PA 51

9x17mm

(.380 ACP)

I have included this pistol among the weapons to be tested for just two reasons: a small quantity of them were purchased by the U.S. Navy after World War I and supposedly issued, and, more interesting to me, the weapon was used by General George Patton.

General Patton was knowledgeable about weapons. Although I do not think he was a handgun expert, he obviously thought about the matter, followed trends, and was willing to put his money where his mouth was on the subject. In contrast to some of his fellow general officers who appeared to be able to move vast armies but who took no interest in actual fighting, General Patton was willing to go out and do personal battle. From 1916, as a young junior officer in Mexico, until his death in Germany in 1945, General Patton carried handguns of one type or another. He was most famous for his pair of belt guns, a 5 1/2-inch .45 Colt SAA and a 3 1/2-inch .357 Magnum Smith &

SPECIFICATIONS

Name: Remington PA 51

Caliber: .380 ACP

Weight: 1.5 lbs.

Length: 6 5/8 in.

Feed: Single-column box magazine

Operation: Delayed blowback

Sights: Blade/notch

Muzzle velocity: NA

Manufacturer: Remington

Status: Obsolete

Wesson, but he also carried a pocket pistol of some type tucked into his waistband. The Colt Model 08 .32/.380 ACP was available, but General Patton wanted something better. He selected a Remington PA 51 .380 because it was the thinnest pistol of its type at the time (1944). Although out of production since 1934, a Remington PA 51 was located, checked, and shipped over to accompany the general throughout the invasion of mainland Europe.

On the formal range, I found the sights narrow, small, and dark. The rear sight notch was very difficult to use, and my groups suffered as a consequence. On the cinema range, the same sight problems made it difficult to index rapidly and to triangulate. The pitch of the grip made it quite handy on the range, aiding instinctive work. I found the thinness of the weapon quite charming as well.

The hammerless design is less than ideal, however, and I would not carry it cocked and locked. A magazine safety would be nice, especially with such a hammerless design.

Although there are clearly superior choices today, in 1934 when the Remington PA 51 was last produced, it had many fine points to recommend it. As with many pocket pistols produced in the 1920s and 1930s, they are frequently found in all sorts of odd places and in the hands of people who normally are not gun people. If you locate a PA 51 in your travels, consider yourself lucky. I would agree with General Patton that, given a choice between the PA 51 and an 08 Colt, the Remington would win every time.

The author test-firing the Remington PA 51.

The 50-foot test target with the Remington PA 51.

Smith & Wesson Combat Masterpiece M15

9x29mmR

(.38 Special)

This is a modification of the target shooter's K-38 revolver introduced in October 1949 by Smith & Wesson. Almost from the start, U.S Marine and Navy aviation units bought significant quantities of this weapon. Later, during the Vietnam War, it became the standard U.S. Air Force issue weapon. Basically it is a refined version of the 4-inch Military and Police (M&P) model.

It has the new short-action style. The tapered 4-inch barrel has a rib made integral to it as with the postwar .38 model, except that the M15's is narrower, resulting in a lighter weapon. The M15 is available in both 2- and 4-inch barrel lengths in the U.S. military, although the latter is by far the most common. The pistol lacks the encased ejector-rod housing of the M19 Combat Magnum, but this may actually be better for a military weapon because it doesn't get clogged with mud. Most people do find it less elegant, of course.

SPECIFICATIONS

Name: S&W Combat Masterpiece M15

Caliber: .38 Special

Weight: 2.1 lbs.

Length: 9.10 in.

Feed: 6-round

Operation: Revolver, DA

Sights: Front blade; rear adjustable

Muzzle velocity: NA

Manufacturer: Smith & Wesson

Status: Current production in service

The M15 offers all the good features of the M&P model, plus the more desirable muzzle-forward balance that the ribbed barrel gives it and better sights. The front sight is of good size, has ramped-in construction, and allows good indexing. It is dark on the cinema range, as all black sights are, but a little white paint remedies this. The rear sight offers a much better picture than what is found on the fixed-sight M&P and helps with triangulation. I have always found that it needs to be widened by about 35 to 50 percent in width for maximum speed in target acquisition and painted white, but, even as is, it is higher on the frame and offers better contrast for quick indexing. Of course, the adjustable sights allow you to rapidly correct the sights to an individual shooter's preference, which aids the better shooter. Less-knowledgeable shooters think they can get better results by moving their sights all the time, and with the M15 they'll have a weapon never properly sighted (but they would get equally bad results either way).

The M15 was used in the U.S. military with the M41 cartridge, a 130-grain, round-nosed, full-metal-jacketed bullet at about 900 fps—obviously no better than the .38 Colt revolver that was so inadequate in the Philippine insurrection of 1892. Similar horror stories were common during the wars in Korea and Vietnam. The more knowledgeable either loaded their weapons with commercial anti-personnel ammunition (a court-martial offense but one, in my experience, rarely ever applied) or did as my friend Leroy Thompson did while serving in the air force in Vietnam. His specialized operation group had M15 pistols authorized. Each man drew the M15, then stored them, and carried M1911A1 or P-35 semiautomatic pistols instead. I well remember Jeff Cooper's recommendation to a young pilot writing about what weapon to carry for his squadron that was issued M15s. Cooper recommended Colt Commanders in .45 ACP, with four spare magazines, which is a much better fighting handgun.

Still the M15 shoots like a target revolver on the formal range and is excellent on the cinema range. I find that the lack of an adapter on the grip causes the weapon to shift in my hand in rapid-fire strings, but it is still quite fast on the cinema range.

True, my years of shooting Smith & Wesson revolvers probably improve my results, yet I must say I can make shots on the cinema range with a K-frame Smith & Wesson 4-inch (any K-frame, M15, M19, M66, etc.) that I cannot make with any other handgun from the same era. In my hands, it is a responsive weapon.

The Smith & Wesson M15 .38 Special (right view).

The Smith & Wesson M15 .38 Special (left view).

The 50-foot test target shot with the Smith & Wesson M15 .38 Special.

Recoil is low because of the steel frame and the power level of the .38 Special cartridge. The double-action pull is generally very good and occasionally great, especially with a well-broken-in specimen. The single-action trigger is always good.

Capacity in any revolver is always going to be lower than with a semiautomatic pistol, and obviously maintenance is more difficult since it takes a lot more skill for an armorer to fix a revolver than merely replace parts that are interchangeable in a semiautomatic. However, if you can accept the idea that you will use your weapon only to defend yourself from one or two enemy soldiers at a time and are willing to pay the cost, the combat revolver has a lot to offer. The biggest drawback to the M15 is not even the cartridge (.38 Special); it is the M41 loading. With the development of soft body armor and ammunition pouches containing steel magazines full of steel-cased ammunition, something better is needed than a 130-grain, round-nosed, full-metal-jacketed bullet at 900 fps. A round such as the THV will put the M15 (or similar weapon) right back into the thick of things.

A U.S. military advisor early in the Vietnam era armed with a Smith & Wesson M15.

U.S. aviator during the Vietnam War armed with the Smith & Wesson M15.

Smith & Wesson M10

The Smith & Wesson Military and Police revolver was designed in 1899. Originally, it was viewed as a more powerful weapon than the typical .38 revolvers of its day, because it was designed to accept a 21-grain, black-powder load rather than the 18-grain load used in the .38 Colt revolver that had proven to be such a failure in the Philippines.

Although there were no doubt Military and Police-style revolvers used in World War I (I have read of them being sold to National Guard units during the period), the weapon came into its own as a military weapon during World War II. Originally, the British government took tens of thousands of the weapons to help arm its troops. They were taken to make up for the deposit the British had placed with Smith & Wesson when the company was producing its 9mm light rifle. That rifle was a failure, but the .38 Smith & Wesson-calibered Military and Police revolver was a great success in British circles. The revolvers were sold in blue and later parkerized finishes in 6-, 5-, and 4-inch lengths. Most people who used them preferred them to the rougher Enfield revolvers, although I believe the Enfield is a superior combat arm because it is faster to reload and tougher in the field. But the finish on the early pre-1942 Smith & Wessons was beautiful, as was the trigger pull on all pre-1945 Smith & Wessons.

After the United States got into the war, Smith & Wesson began producing the Military and Police model in .38 Special with parkerized finishes with either 4- or 2-inch barrels. Most of these weapons were destined to go to military plant guards or the FBI, but some were actually issued to pilots in the U.S. Navy and Coast Guard. Many nonshooters of the day preferred the lighter, smaller .38 revolver to the .45 Government Model 1911Al, and, accordingly, there are numerous

SPECIFICATIONS

Name: S&W M10

Caliber: .38

Weight: 1.9 lbs.

Length: 9.25 in.

Feed: 6-round

Operation: Revolver, DA

Sights: Fixed, blade/notch

Muzzle velocity: N/A

Manufacturer: Smith & Wesson

Status: Current production, in service

reports of these being stolen from stores by light-fingered U.S. troops.

When the war was over, many of these revolvers were surplused off, and they were a common feature in U.S. gun magazines in the 1950s and early 1960s. Most were British surplus (Lee Harvey Oswald was carrying a British surplus Military and Police rechambered to .38 Special when he was captured after allegedly killing President Kennedy). But while the government was surplusing off wartime Military and Police revolvers, it continued to purchase new ones. Again, they went mainly to aviation units, security guards, and (where I encountered them) to the Criminal Investigative Command. Apparently, the powers that be thought that CID agents were like FBI agents and should have a 2-inch-barrel revolver available. to them. Typical of the frequent lack of understanding, the CID command thought short-barrel revolvers were easier to conceal than 4-inch models, when actually it is the width and height that are key concerns.

When I first got involved with CID operations, we had Colt Detective Special 2-inch revolvers. Around 1975, they were withdrawn and 2-inch M10 Smith & Wesson revolvers (the new postwar short-action model) were substituted. The Smith & Wessons were bigger and heavier than the Colts, and, frankly, I did not like them. I would have much preferred a 4-inch model of the same pistol.

The 2-inch Military and Police model has always seemed to me to be the worst of all worlds. It is heavy for its size, big, light in caliber, and have a short barrel which makes it difficult to shoot accurately. I have always preferred the J-frame Smith & Wesson in a 2-inch-barrel .38.

Although I had this experience with the 2-inch Military and Police prior to testing it for this book, I cannot really say that I had done too much work with it. After having been given one, I had placed it in the weapon safe and carried my lightweight Commander .45 and my M38 2-inch .38.

I must say I was pleasantly surprised on the formal range with the M10. The trigger was good, and the sights large enough to be easily seen and kept on target. Accuracy was quite good. Recoil was mild due to the weight of the pistol and good size grip, in comparison to what I am used to from +P loads in my air-weight J-frame.

It was even better on the cinema range. The wide front sight coupled with the rear notch allows quick indexing in dim light. Painted white, it would even be better. The smooth double-action trigger allowed for quick firing on multiple shots. The light recoil allowed a rapid repeat follow-up shot. Without an adapter, the grips tended to shift in the hand, but they are quite adequate with a simple adapter installed.

The ejection rod is too short to clear the cases if they get stuck, because the 2-inch barrel does not have a full-length ejection rod, as is the case with a 3-inch or longer barrel. Still, the low-pressure .38 Special loads do not expand the brass severely, and a quick strike on the rod with the palm of the hand clears all but the most stubborn cases.

The hammer spur does catch on clothing occasionally, but this could be fixed in a few minutes—although in all honesty I never saw one so modified while I was on active duty. Most soldiers would be too fearful of the consequences of doing so.

On the cinema range, the 2-inch Military and Police-style Smith & Wesson was quite handy. Reflecting back on the formal range test and then the cinema range, I started thinking that perhaps my evaluation 20-plus years ago had been in error; maybe this was a good combat revolver. But then I looked down at my hand and realized why I did not carry one then and why I would not carry one now: the Military and Police 2-inch is simply too big and heavy to be comfortably carried in the pocket of your field jacket, and the positives of a good-size grip and weight don't compensate for these shortcomings. A weapon that does what the Military and Police does in 2-inch format should be no bigger than a 2-inch J-frame and weigh 19 ounces or less. Such a weapon may be more difficult to shoot, but it can go to the field where the Military and Police would be left "in the rear with the gear."

The author test-firing the Smith & Wesson M10 .38 Special (above left).

The 50-foot test target shot with the M10 (above).

Right side of the Smith & Wesson M10 2-inch .38 Special (left).

Left side of the Smith & Wesson M10 .38 Special (below left).

Colt Detective Special
9x29mmR
(.38 Special)

At one time, this weapon was the standard-issue weapon among U.S. Army Criminal Investigative Command special agents, and this is where I first encountered it. Because it was in the inventory, it also trickled into other units as well. Further, it was a standard-production handgun, and many an individual soldier purchased it and carried it off to war.

With the addition of the hammer shroud, which was a common factory part, this weapon was ideal for carry in military parkas and coat pockets. Even today, if loaded with THV ammo, it would be quite suitable for use against vests found on standard Western infantry troops.

The weapon is rather difficult to shoot because of the stacking of the trigger, unlike with the Smith & Wesson M10 2-inch revolver. However, the Colt has better sights than the standard 2-inch J-frame Smith & Wesson, and the Colt is smaller than the K-frame M10 Smith & Wesson.

SPECIFICATIONS

Name: Colt Detective Special

Caliber: .38 Special

Weight: 1.5 lbs.

Length: 7 in.

Feed: Revolver

Operation: DA

Sights: Blade/notch

Muzzle velocity: NA

Manufacturer: Colt

Status: Limited current production, in service with CID units

I find this to be a very good service weapon for line troops and much more deadly than a bayonet. Time spent in bayonet training (except to teach aggressiveness) is better spent learning how to shoot pistols like this model. A short-barrel .38 with proper ammunition will allow a soldier to stop five or six enemy soldiers, whereas he would be lucky to get one with a bayonet. Although I do not believe this weapon is as good as a Smith & Wesson M940 2-inch because of the latter's stainless steel construction, for real-world infantry troops, it is a lot better than an M1911A1 or similar-size weapon. For those who do not carry a rifle, the Government Model-size pistol (whether a Glock 17 or something else) is fine. For those who do carry a rifle—along with a shovel, extra ammo, grenades, radio battery, etc., as is common with real troops—a revolver such as this, tucked in some out-of-the-way place, ready to be grabbed when your rifle is unavailable, is worth its weight in gold even if you never use it.

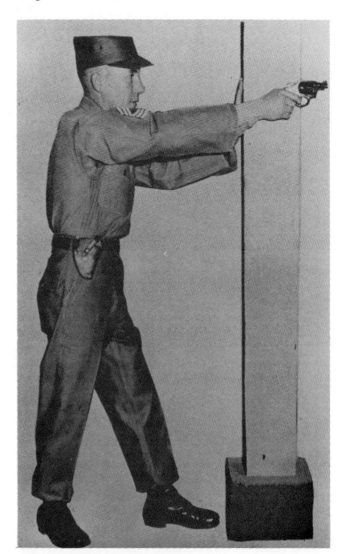

U.S. Army CID agent armed with a Colt Detective Special 2-inch, which formerly was standard issue.

The author test-firing the Colt Detective Special 2-inch revolver.

Early U.S. advisor to Vietnam and Vietnamese soldier, who is armed with the Colt Official Police handgun.

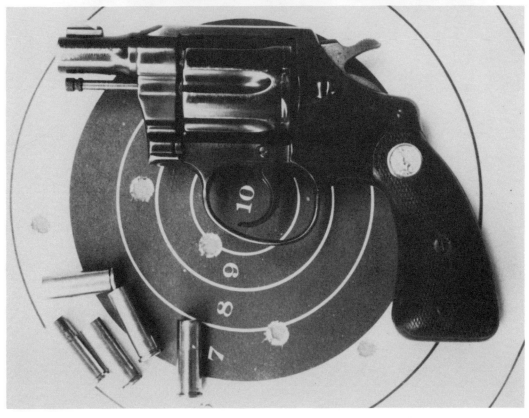

The Colt Detective Special 2-inch with 50-foot test target.

Smith & Wesson M60

9x29mmR
(.38 Special)

This was "the pistol" during the Vietnam War. But it was not because of its caliber, design, or quality of construction, but rather its material. Announced in 1965, the M60 was the first handgun made entirely out of stainless steel. Everybody who knew anything about handguns and anticipated going to Vietnam wanted one. I immediately started to look for one, but I never could find one. Instead, I had to settle for a 3-inch nickel M36 Chief's Special Smith & Wesson, and I was happy to get it.

Small-frame Smith & Wessons were hard to find between 1965 and 1975. I remember the factory-list price of an M60 being $85 and people happily buying them for $200. Leroy Thompson reports that among his operational group in Vietnam, an M60 could be swapped for anything you really wanted: booze, guns, or women. He sold his M60 for $500 on the way back home in 1968.

SPECIFICATIONS

Name: S&W M60

Caliber: .38 Special

Weight: 1.2 lbs.

Length: 6.5 in.

Feed: Revolver

Operation: DA

Sights: Blade/notch

Muzzle velocity: NA

Manufacturer: Smith & Wesson

Status: Current production

Small pocket pistols are typically carried close to the body and are subject to heat and humidity extremes. For years, blue-finished handguns have been turning brown on people. In Vietnam, they often turned brown literally overnight. Nickel-plated guns were fine as long as the nickel was intact, but as soon as it chipped or flaked, the whole finish was under attack. Stainless steel altered this whole equation. People who never would have accepted a bright pistol before flocked to the M60.

The M60 suffers from the same drawback of all J-framed Smith & Wesson revolvers: the sights. The front sights are too narrow (except in the most recently released models that finally have gone for 1/8-inch sights), and the rear-sight notch is narrow and shallow. The short barrel magnifies any sighting error, also making it difficult to get good groups on a formal range. The trigger pull on single-action is not bad, but the coil springs and short hammer fall usually produce a very heavy, stiff double-action pull. The weapons are light and hence recoil a bit; the barrels are short, so muzzle blast is high; the grips are small, allowing the weapon to shift in the hand. I cannot shoot one accurately without a grip adapter installed, because the weapon shifts excessively in my hand.

Having enumerated all these bad things about the M60, I must say that I believe it to be one of the finest handguns used by *real* fighting soldiers that is still available. Clearly, it was the best weapon for the company-grade infantry officer in Vietnam. When you carry a rifle, spare magazines, and grenades and have to walk long distances on patrol, you want a handgun that can be slipped into the upper left breast pocket of your fatigues, out of the way. Only if your rifle is unavailable will you use your handgun. But when you need it, it needs to be available immediately. The M60 gave this to the infantryman of the Vietnam War.

In the years since leaving the infantry, I have reconsidered many times my conclusion about the M60. In writing this book, I have reviewed most of the military handguns in use during the twentieth century, and I think my 1966 decision was correct. Today, my choice might be an M638/642 or an M940 with THV ammunition, but those weapons and the ammunition are "kissing cousins" of the M60 with 110-grain SuperVel ammunition I sought in 1966. In my estimation, the J-frame stainless steel (fully or in combination with alloy) in .38 Special or in 9x19mm with a 2-inch barrel is the single best military handgun for any fighting soldier.

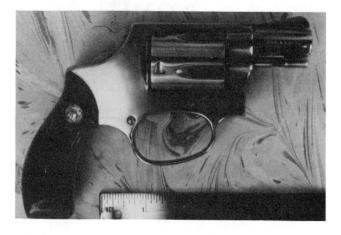

The Smith & Wesson M60 .38 Special was the most desirable gun in Vietnam.

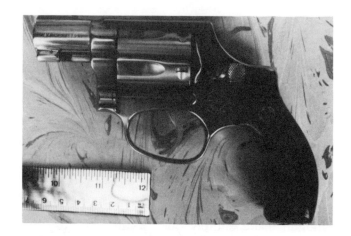

The left side of the Smith & Wesson M60.

The 50-foot test target shot with the Smith & Wesson M60 2-inch .38 Special.

Smith & Wesson Model 39

9x19mm
(9mm Para)

The Smith & Wesson Model 39 9mm pistol really started the movement in the United States to double-action automatics. At the conclusion of World War II, the U.S. military, having been very impressed with the 9mm P.38 Walther, put out requests for a new 9mm pistol. Smith & Wesson responded with two models, the double-action Model 39 and the single-action Model 44. Although it appeared that the U.S. Army would adopt the Smith & Wesson, after much political interference, it turned out that the U.S. military stayed with the .45.

Smith & Wesson turned to the civilian market to recoup its research and development expenses. Police departments throughout the United States adopted the Model 39, beginning with the Warwick, Rhode Island, Police Department in 1961. Soon, a few small police departments in California adopted the Model 39, followed in the late 1960s by the Illi-nois State Police. From that point on, double-action 9mm pistols were the wave of the future. For those of you who are latecomers to the field, there were not that many double-action .45-caliber automatics around until the late 1960s—in fact, there were none to my recollection. Thus, a double-action U.S.-manufactured (by a reputable manufacturer) semiautomatic pistol was very desirable for police purposes.

I include the Model 39 in this study of combat handguns because, during the Vietnam War, many U.S. soldiers carried it, and a small quantity was purchased by the U.S. Air Force. The U.S. Navy SEAL teams also carried what it turned out to be the Model 59 double action. But, mainly, the Model 39 was brought along as a personal handgun.

The double-action pull on the Model 39 is quite heavy and is a little too long for most people's trigger fingers. Single action tends to have a lot of slack and overtravel, but it is not as heavy.

SPECIFICATIONS

Name: S&W Model 39

Caliber: 9mm Parabellum

Weight: 26 oz.

Length: 7.4 in.

Feed: Single column box

Operation: Recoil

Sights: Blade and adjustable rear

Muzzle velocity: NA

Manufacturer: Smith & Wesson

Status: Out of production but still in common use

This weapon, though possibly too large for its power, is really quite handy. Its aluminum frame weighs in at 26 ounces, so it is convenient to carry. It is also reliable; all you do is pull the trigger for your first shot, and no further action is required (one reason why many police agencies like it so well). Similarly, for noninfantry types, such as clerks or helicopters pilots, it is convenient to carry a weapon loaded without having to worry about safeties—simply pull out the gun, pull the trigger, and off it goes.

The main problem with the Model 39 in the early years was loads. Only 9mm, full-metal-jacketed loads existed until recently. Nowadays, of course, we have better ammunition available in 9mm, namely the BAT round, THV round, and a variety of soft-nose and specialty ammunition. For military purposes, of course, all the soft-nose specialty ammunition is prohibited. One big drawback of the Model 39 series is the fact that unlike the Czech Model 75, it is a nonselectable double-action pistol, which means that after you fire the first couple of rounds, you have a cocked pistol. You have two options. One, you have to drop the safety, causing the hammer to drop, then push the safety back up in place so that you can carry the weapon with the hammer down, ready to be fired again, which causes the trigger to have a long, double-action trigger pull. Two, you carry it with the hammer cocked and no safety on and run around with a cocked, single-action pull available. Obviously this latter method is dangerous and should be avoided if at all possible. It would be nicer if the Model 39 had a selective double-action pull like that of the CZ 75 so this problem could be avoided.

In the United States, Smith & Wesson autoloaders are common police weapons, although their dominance is now being threatened by overseas designs. The SIGs have become common and Beretta double-action, double-column, 9mm self-loaders seem to have developed quite a following. The Smith & Wesson Model 39 series is still quite good and fairly inexpensive. Single-column versions are generally better than the double-column Model 59 because the grips are easier to handle for most people with average-size hands.

Accuracy on Smith & Wesson autoloaders has always varied widely. Some of them shoot spectacularly well: I have seen one Model 39 that shot 1 1/4 inches at 25 yards with good ammunition. I have seen others that could not shoot into 8 inches. Barrel fittings seem to be the critical concern for all Smith & Wesson automatics, and a properly fit barrel goes a long way toward solving the problem.

When I was employed by the federal government, I used the Model 59 for quite a while, and it broke to pieces after thousands of rounds were fired through it over an 18-month period. They are not the same quality as the SIG P210 by any means. You can't shoot 250,000 rounds through them, but they seem to be reliable for average military or police use: the kind of pistol that you are going to shoot 500 rounds a year through for the next 20 years. They're not designed as a match-shooter's gun, but rather as a serious weapon for combat or military use. Although I prefer the Glock 17 in 9mm or the SIG P225 in 9mm, the Model 39 has to rank up there with one of the top 9mm double-action pistols in the world.

Right side of Smith & Wesson M39 9x19mm pistol.

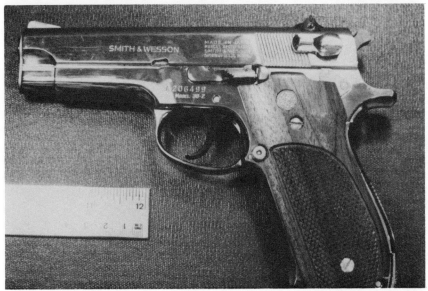

Left side of Smith & Wesson M39 9x19mm pistol.

The 50-foot test target shot with the Smith & Wesson M39.

ASP

9x19mm
(9mm Para)

The ASP 9mm was a modification of the Model 39 Smith & Wesson 9mm autoloader. Designed originally in the late 1960s by Paris Theodore of Seventrees Limited fame, this modification—and not the original Model 39—became quite well known and was favored by many recognized pistol shooters of the day. I have included it in this evaluation of combat handguns only because it was used in the tail end of the Vietnam War by U.S. intelligence and security agency personnel. For example, the former chief of security at the American Embassy in Saigon carried an ASP pistol. Although technically speaking it has not been adopted as a war weapon by any military force, it is an interesting variation.

The size and weight on the ASP are handy. The trigger mechanism is the same as with a standard Model 39 Smith & Wesson in that the single-action trigger has a lot of overtravel, and

SPECIFICATIONS	
Name:	ASP
Caliber:	9mm Parabellum
Weight:	1.5 lbs.
Length:	6.75 in.
Feed:	Box magazine single column
Operation:	Recoil
Sights:	Fixed 1-piece unique guttersnipe-style sight design
Muzzle velocity:	NA
Manufacturer:	S&W and modified by Armament Systems & Procedures
Status:	No longer available

the double-action trigger has a tendency to stack. Additionally, as with all Smith & Wessons, the hammer fall is moderate, and with hard primer submachine gun ammunition, you sometimes have to fire it or click it two or three times before it will go off. But if you are using modern, pistol-designed, 9mm ammunition, this is generally not a problem. The Teflon finish applied after modification provides a reliable, smooth, dark-finish weapon with no flash. I think it is a promising finish for combat pistols and is certainly something to be looked into. I have probably put more than 2,500 rounds through my ASP. The first time I got it back from the remanufacturer, I found that it had not been properly heat treated. It shot loose and had to be sent back to be completely rebuilt. Apparently, when the Teflon finish is applied, it has to be baked on the weapon, which has a tendency to draw the temper considerably. You have to watch out who applies it.

The cut-away sides of the magazine and the translucent grips showing how many rounds are left over are features that would enhance any military pistol.

Probably the most characteristic feature of the ASP system is the Gutter-Snipe sights, patented by Paris Theodore. Much has been said about how wonderful his sighs are: how fast they are, etc. Well, maybe they are for others, but for me, they were almost impossible to see and extremely difficult to use at a distance. If you shoot in practical speed situations—during which your eye is concentrated on the point on the target—you basically throw your bullets from your pistol to that point by indexing your sights, putting the front sight on the bottom of your field of vision, picking it up subconsciously, and indexing it against the rear sight on triangular basis. The ASP Gutter-Snipe sights will not work this way. If you use the sights, you cover up your target with them, making it difficult to see, and follow the target during cinema range situations. On the formal target range at 50 feet, I ended up with groups roughly 6 3/8 inches in size because holding my elevations and windage was very difficult. The common excuse to explain poor performance is that the Gutter-Snipe sights are designed for high-speed situa-

The 50-foot test target shot with the ASP 9x19mm handgun. The pistol's sights make good groups very difficult to obtain. Conventional sights result in better groups.

tions. But they were not any faster on the cinema range. Further, accuracy suffered considerably on the cinema range because of the way the sights are designed. If my pistol was modified again to this size, I am afraid that the Gutter-Snipe sights would have to go. I would just as soon have a good set of conventional sights, rounded so as to avoid snagging. So a good size, good finish, interesting features, so far as the cut-away magazine, and hammer are offset by the sights, making this pistol less than desirable as a combat weapon.

Right side of ASP
9x19mm handgun.

Left side of ASP
9x19mm handgun.

Heckler & Koch
P9S (SEAL)
9x19mm
(9mm Para)

The Heckler & Koch P95 handgun is now out of production, but it was available in both 9x19mm and .45 ACP. It had a plastic frame before the Glock 17 did, but it did not get any hostile press. As a consequence, it became known only to firearms enthusiasts.

The weapon came with fixed sights and short barrel or adjustable sights and a longer barrel. The weapon had a double-action trigger system, allowing the weapon to be carried hammer (striker) down and then fired by pulling the trigger. Alternatively, it could be cocked by pushing the cocking lever and fired in the single-action mode.

A few years back, the U.S. Navy SEALs, as part of their ongoing weapons program, had a number of P95 pistols made specifically for them. The pistols were specialized in that they had the double-action feature locked out. They used the target trigger system, which also allowed the

SPECIFICATIONS

Name: H&K P95

Caliber: 9mm & .45 ACP

Weight: 2 lbs.

Length: 9.1 in.

Feed: Single-column box magazine

Operation: Roller lock

Sights: Adjustable blade front and adjustable rear

Muzzle velocity: NA

Manufacturer: Heckler & Koch

Status: Obsolete

weapon to be adjusted to fire a three-shot burst if desired. With a single-column magazine of nine rounds, this feature is not very useful in my judgment.

As noted, the sights were unique in that they fitted the adjustable rear and then added a very high sight to the normal slide. This was done because these pistols were typically fitted with a suppressor. The P95 was admirably suited to this use because the barrel was fixed and the locking system was the H&K roller-lock rather than the typical Browning recoil-locking system. By virtue of that barrel attachment/locking system, the suppressor could be fitted and the weapon be sighted in properly—and still avoid an off-center suppressor. Used to take out guard dogs and isolated sentries, this weapon must have been very effective. It was too specialized for general military use, however, and was not frequently used.

On the testing range, the proof of the design

showed in the 1 1/4-inch group at 50 feet. This is in comparison to the standard 2-inch group at the same distance with the Smith & Wesson M19. The P95 has an excellent trigger pull. I did not experiment with the burst-fire capability because the limited magazine capacity rendered that feature worthless in my judgment, and I did not want to ruin the excellent single-action trigger pull. To be effective, I think a burst-fire pistol needs at least a 20-round magazine. Three-shot bursts are not uncommon in combat situations, and you really should be able to fire a weapon four times without reloading, which is not feasible with a nine-shot pistol.

The sights permit rapid pickup in the darkened cinema range, as well as rapid indexing. Unfortunately the single-action trigger system makes it difficult to handle rapidly if it is not cocked to start with. If it is cocked, you must flip off the safety. Disengaging the safety is not as quick as with the simple down-stroke method of the Colt Government Model design, and reapplying it is also slower. I suppose for the purposes designed, this delay was not a big drawback for the SEALs.

The author test-firing the Heckler & Koch P95 9x19mm.

Right side of the Heckler & Koch P95 9x19mm.

Left side of the Heckler & Koch P95 9x19mm.

Beretta M92 (M9/M10)

9x19mm
(9mm Para)

After World War II, the U.S. Army undertook trials to replace the .45 Government Model with a 9mm handgun in line with NATO standards. Ultimately, the trials led to the development of the Smith & Wesson M39 and the Colt Commander pistol. But by 1954, the military decided the services had enough handguns and dropped the whole project.

Matters stayed this way until 1982, when, after much testing, a lot of false starts, and much debate, the U.S. military adopted the M92 Beretta. After lengthy trials, this pistol was selected over the SIG P226 only because of its lower price per gun. After both weapons were judged to be satisfactory, the intelligent money was on the SIG, but the Beretta won out. Many people believed it was because the United States wanted to install cruise missiles in Italy at the time, and it was a political concession. Perhaps, but since its adoption by the U.S. military, it has been adopted by numerous po-

SPECIFICATIONS

Name: Beretta M92

Caliber: 9mm Parabellum

Weight: 2.5 lbs.

Length: 8.54 in.

Feed: 15-round, in-line, detachable box mag.

Operation: Recoil; semiauto

Sights: Front blade w/ slide; rear notched bar

Muzzle velocity: 1,268 fps

Manufacturer: Pietro Beretta SpA

Status: Current production & use

lice forces and other military agencies in other parts of the world who figured (quite correctly) that any weapon rugged enough to comply with U.S. military requirements would be satisfactory for their militaries also.

As a military handgun for infantry troops, the Beretta M9 or M10 (which are the same except for a slide-catching device installed after the SEALs broke the slides on their M92 pistols from heavy firing), leaves a lot to be desired. Although the weapon's sights are easy to use on both the formal and cinema ranges and did offer a nice contrasting color, larger sights would improve the ability to index rapidly on the cinema range. The pistol weighs as much and is just as big as (and actually thicker than) the Colt Government Model it replaced. The double-action trigger mechanism is not selectable, so you run into all the typical problems, namely people running with them cocked and no safety on after a shooting incident when they refuse to drop the hammer and start over with a heavy, double-action pull. Interestingly enough, at a time when the U.S. military is putting more females and smaller-size soldiers into the service, the grip size is bigger and the double-action trigger system makes it harder for smaller people to handle than the single-action trigger style of the Colt Government Model.

The M92 has a light recoil, which is to be expected from a pistol that weighs 40 ounces and shoots only 9x19mm cartridges. For some reason, in my test on the cinema range, the flash with the Beretta was greater than it was with a CZ 75, even though I used the same ammunition. Perhaps it was the difference in locking systems. The Beretta uses the Walther-type lock, which generally makes for a wide pistol.

The Beretta M9/M10 has now been in service for a number of years. Some problems developed early with broken slides, and nobody was ever able to say why they broke. At one point, the U.S. military issued books to write down how many rounds were fired in a given pistol and replaced slides every 1,000 rounds. On the other hand, everybody admits that the best feature on the Beretta M9/M10 is its reliability—at least until the slide broke without warning.

Oddly enough, the weapon lacks a magazine safety but does offer a loaded-cartridge indicator. Although available in stainless steel for police and civilian sales, the military chose the steel guns—an odd selection. The military issued it with two spare 15-round magazines; I prefer one spare 20-shot magazine myself. But then, everyone I have spoken with in the military who carries the Beretta tells me they are limited to having just five rounds in the magazine or carrying it in condition three, which totally negates the benefit of getting a double-action autoloader. Last, the United States uses a ball 9x19mm round except for certain elite units, and this round offers none of the good points of the 9x19mm (such as armor penetration) and is a poor stopper on nonarmored subjects in contrast to the .45 ACP round.

In short, the M92 is an acceptable handgun but nothing great. I have really tried to like this pistol because I think everyone should be familiar with his country's service handgun, but I have to admit that I cannot get myself to love or even like it. Originally, I thought I was merely being an old mossback. I really like Government Model .45 autoloaders. I have carried them in the field while in the service, worn one as a U.S. Marshal, won money with them, killed animals with them, and taken men at gunpoint when only my .45 prevented me from getting shot by a sawed-off shotgun. Perhaps I was being like a

The author test-firing the Beretta M92 (M9/M10).

boy who liked the Colt SAA .45 with black-powder loads when the M1911 first came in. But after buying a pair of M9/M10 pistols and really working with them, I have concluded that they offer nothing that cannot be obtained elsewhere in a better package.

Beretta M92 (M9/M10)

Right side of the Beretta M92 (M9/M10).

Left side of the Beretta M92 (M9/M10).

The Beretta M92 (M9/10) with the 50-foot test target shot with it.

SIG P226

9x19mm
(9mm Para)

This is the pistol the U.S. military forces actually wanted when they adopted a 9x19mm pistol. Many of the specifications are the same, except that it has a decocking lever only and no safety. Many elite military units have used this pistol, the most famous being the Navy SEALS. After they broke the M92 (M9/10) repeatedly, they refused delivery of any more Beretta pistols and bought SIG P226 pistols instead. This pistol is also used by many federal law enforcement agencies. For instance, it was the first autoloader adopted by the U.S. Secret Service for its executive protection teams. It is general issue for FBI S.W.A.T. teams, other than the national team, which uses the P-35.

The SIG P226 has also just been adopted by the British army to replace its aging stock of P-35 pistols. British troops use the shortened P228 for some duty functions, but the standard P226 is the

SPECIFICATIONS

Name: SIG P226

Caliber: 9mm Parabellum

Weight: 1.5 lbs.

Length: 7.72 in.

Feed: 15-round, detachable box mag.

Operation: Short recoil; SL; SA or DA

Sights: Front blade; rear square notch

Muzzle velocity: 1,138 fps

Manufacturer: Schweizerische Industrie-Gesellschaft

Status: Current production

general-issue weapon now.

The SIG P226 is an extremely reliable, highly accurate weapon. Single-action pull on the SIG P226 is always good. The double-action pull is quite stiff with a lot of stacking. Fortunately, a shorter trigger can be installed, which improves leverage and makes the pull seem easier. The SIG pistols of any style are always quite accurate. They vary from excellent to outstanding in my experience.

The nonselectable double-action pull is a shortcoming on this pistol. The hammer drop is located at the widest point of the grip, which causes you to have to shift the weapon in your hand and thereby lose your grip to decock the weapon.

The front sight is very good for fast pickup on the cinema range, as is the rear sight. The extra capacity of the magazine is a feature that many users like, but I think the resulting bulkiness is too high a price to pay. I prefer the SIG P225 to the P226, but I must say I am in a distinct minority on this issue. Many people like the lack of a safety device on the SIG P220/5/6/8 pistols, and for a military gun, this is acceptable. However, on a civilian pistol a safety is desirable because it may avoid household accidents and weapon-snatch problems. I would like to see a magazine safety installed on a mili-

tary handgun, and I think its lack on the SIG series is a serious drawback.

The pistol is too thick for my tastes, but it is otherwise as good as the P225. I would agree with the SEALs: if you can, pick the SIG over the Beretta M9/10. You can't go wrong by selecting this pistol if you are looking for a full-size battle pistol.

The author firing the SIG P226 9x19mm pistol (right).

Right side of the SIG P226 9x19mm pistol (below right).

Below, a comparison of left side views of the SIG P226 (top) and P220 (bottom).

A comparison of the slides of the P220 (left) and the P226 (right).

The P226 9x19mm with 50-foot test target.

SIG P228 (M11)

9x19mm
(9mm Para)

As I stated in the last section, the U.S. military really wanted the SIG P226 when it sought a 9x19mm pistol, but got stuck with the Beretta M9/M10. Well, some military units got around adopting the Beretta by opting for the SIG P226 anyway, rebuilding old Government Models (as did Delta), or buying Glock 21s, as did the Marine Corps.

After the M9/M10 was adopted, all of a sudden it dawned on some people that this was a *big* pistol. There were many different pistols in the U.S. military inventory simply because big pistols don't work for every situation. But the military had spent a lot of money convincing Congress they that it needed one caliber to rationalize ammunition control, 9x19mm. As soon as the call went out for a smaller pistol for women, criminal investigators, and others, the P228 was developed.

Now, anyone who knows anything about

SPECIFICATIONS

Name: SIG P228 (M11)

Caliber: 9mm Parabellum

Weight: 1.8 lbs.

Length: 7.08 in.

Feed: Double-column box magazine

Operation: Recoil

Sights: Blade/notch adjustable by drifting

Muzzle velocity: NA

Manufacturer: Schweizerische Industrie-Gesellschaft

Status: Current production

guns knows that it is not the length of the barrel that makes it difficult to conceal a pistol, but rather its width. The P228 was shorter than the M9/M10, but just as wide since it used a double-column magazine. If the military services needed a smaller weapon, you think they would have gone to the single-column version of the M92. Had they done that, the manual of arms training would have been the same. Instead, they adopted the P228, calling it the M11, which meant extra training. Of course, the P228 was merely a shortened version of the pistol they originally wanted, and that was probably the major element in the decision.

The British army made a better decision in adopting the P226 for general use and the P228 for those requiring a smaller weapon. It would have been even better for it to have taken the P225, but high-capacity autoloaders are the rage for those who plan to miss a lot.

The P228 offers the same good sight picture, high accuracy, and trigger system as the P225/6 series. It likewise lacks a magazine safety and is not a selectable-trigger-system autoloader. The grips are quite narrow on the P228, and it is only slightly wider by actual measurement than the P225, although it feels different in your hand. The short trigger can be fitted to the P228, however, unlike the P225. As with any shortened barrel, 9x19mm weapon muzzle width is more pronounced than with a longer barrel, but it is by no means difficult to control. The P228 grip is long enough for my hands (size 9), although ham-fisted shooters may find their hands cramped.

This is a fine handgun; only the thinking behind its adoption seems murky to me. Although I would prefer the P225, I can see how the extra ammunition in the grip might be comforting for many people. Those soldiers in the U.S. and British armies who will carry one into battle can depend upon it to the extent that their ammunition is suitable to the project. It certainly must be ranked in the top dozen battle pistols available today.

The author firing the SIG P228 (M11).

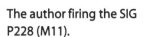

Right side of the SIG P228 9x19mm.

Left side of the SIG P228 9x19mm.

Rear view of P228. Note how thin the grip panels are despite the double-column magazine.

SIG P228 9x19mm with 50-foot test target.

Glock M21

11.43x23mm
(.45 ACP)

When I got my first Glock pistol, a Glock 17, in 1986, like all real Americans I wanted one in .45 ACP caliber. Now after years of thought on the subject of military weapons, I have concluded that the 9mm model with ballistic-vest-penetrating loads is actually a better choice. It appears, however, that some people still think the same as I did in 1986. Hence, the larger-caliber 10mm Glock M20 and .45 ACP M21 were developed. I included the M21 Glock .45 ACP in this volume primarily because of its use by the U.S. Marines.

As is commonly known, the Beretta M92 (or M9/10 as it is known in U.S. military circles) has had some problems. When used by the U.S. Navy SEALs, it suffered slide separation. Depending on whom you speak to, this was because of 1) the SEALs' practice of shooting them from three feet underwater, creating severe pressure problems,

---

SPECIFICATIONS

Name: Glock M21

Caliber: .45 ACP

Weight: 1.6 lbs.

Length: 8.27 in.

Feed: 13-round, detachable box mag.

Operation: Short recoil; self-loading

Sights: Front blade; rear notch

Muzzle velocity: 1,138 fps

Manufacturer: Glock GmbH

Status: Current production

---

2) constant use of excessive pressure specialty loads, or 3) a weak design. Since it had been adopted by a number of police departments and none had reported the problem, and the Italians likewise had never noted this problem, I was inclined to believe it was the SEALs' training program or ammunition that caused the separations. Then in 1991, while I was in Paris, I went around with a member of the French Police R.A.I.D. unit, whose members were authorized to carry the Beretta M92, since the M92G had been adopted by the French gendarme and was being made in France. He reported no one in the unit would use the Beretta because of slide breakage: the same problem reported by the SEALs. Maybe it was the guns!

Finally, things got so bad that the U.S. Navy refused to accept any more M9/10 pistols. The SEALs chose the SIG P226, and the Marines Corps bought the M21 Glock. The Colt Government

Models used by the Marines Corps were apparently worn out, and rebuilding them was not considered economically (or socially) feasible. Additionally, even in the Marine Corps, shooting a lot of rounds is fashionable, and the single-action style of trigger found on the Government Model is apparently too difficult to train the troops with if carried cocked and locked.

My friend Evan Marshall was hired by Colonel Young of the U.S. Marines to train the marines assigned "star-wars" security in the use of the combat submachine gun. Because the submachine gun was not then part of the Marine Corps' inventory, no one was there to teach the weapon when the Corps bought a bunch of MP5 submachine guns. In came the civilian expert. Upon enter-

The author firing the Glock 21 .45 ACP.

ing the secured area, Evan was met by guards carrying cocked-and-locked (condition one) .45 Government Models. Evan commented with approval to Colonel Young about the level of expertise and competence shown by the young marines. Colonel Young told Evan to look in the butts as they passed. When he did so, Evan saw no magazines! Apparently, the guns were carried cocked and locked but with no ammunition. The magazines were carried on the belt. This was the worst of all ways because to load the weapon, the marine had to release the safety, then withdraw the slide, and load the weapon. If the training level at the "star-wars" research facility was this low, you can imagine what it must have been like in a "normal" unit.

The marines chose the Glock M21 instead of the Beretta M9/10 because they could get it in .45 ACP, which they preferred to a 9mm pistol, and yet still have a weapon that was more durable, safer to operate (at least in some minds), and cheaper than new Colt Government Models. Perhaps if the marines had given more thought to

ballistic vest/helmet penetration, they would have taken the 9mm version, but the marines who selected the .45 ACP Glock were familiar with fighting unarmored Vietcong/North Vietnamese Army troops.

The Glock M21 has all the good points of the Glock 17, except for handiness. The Glock M21 is simply too big in the grip for most people, and the slide is too wide for easy concealment. This latter point is not critical for a military handgun, but the grip problem is. A single-column .45 ACP Glock would avoid the problem of grip width, but the cartridge limitation of .45 ACP would still be present. The U.S. Marines' choice of design was excellent, but their preoccupation with the past caused them to get less than the best solution to their problem. With the possibility before them of getting the best military handgun in the world, they blinked and got something excellent for large-handed troops as long as unarmored targets were involved. The Glock 21 is not nearly as good when modern, fully equipped troops are encountered.

Right side of
the Glock 21
.45 ACP.

Left side of
the Glock 21
.45 ACP.

Ammunition Selection, Lanyards, Holsters, and Desirable Features for a Military Handgun

When discussing military handguns and ammunition, we must remember that the military handgun is just like its civilian counterpart: it is really only a projectile launcher. It doesn't matter what the weapon looks like or how it is made in terms of stopping or killing the enemy soldier. It matters only how effective the projectile is when it hits human flesh.

AMMUNITION SELECTION

The ammunition is the key factor in how well the weapon system will perform. With this in mind, we must remember a couple of key points. The projectile must actually hit human flesh to do any damage. If it is stopped or deflected by the typical equipment worn on the battlefield, then it will do no harm to the enemy. This is an important observation since the typical enemy soldier will be wearing web gear containing loaded magazines, canteens, first-aid pouches, grenades, and similar gear, and, in advanced armies, ballistic body armor. People in a civilian environment will rarely be wearing items that are likely to stop or deflect a projectile.

TEST RESULTS OF SHOOTING INTO MILITARY EQUIPMENT

Two AR-15 magazines filled will ammunition shot at 10 feet into a GI magazine pouch:

1. 9mm ball from a Beretta M92 penetrated the first magazine and got caught in the second.
2. 9mm Makarov in RK 59 did not go through even one magazine; it just left a large dent.
3. .45 ACP GI ball in Smith & Wesson M4506 penetrated one magazine and dented the second.
4. Czech AP rounds in Beretta M92 had complete penetration of both magazines, pouches, and a 900-page book; the bullets were found in the ground with the core missing. NOTE: There was no ammo explosion or detonation.

Front view of magazine test target that contained two loaded M16 magazines.

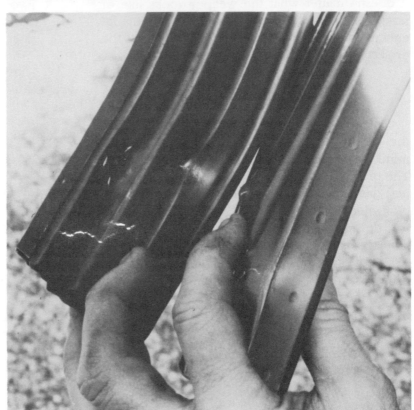

Deep dent, where finger is pointed, in first loaded magazine left by 9mm Makarov ball round. There was no penetration.

After being shot by .45 ACP ball round, one magazine was completely penetrated, and the round entered the second but did not exit (top left).

The entrance hole made with the .45 ACP round (top right).

Exit hole made in the first magazine by .45 ACP round (above left).

A rear view of the second loaded magazine struck by the .45 ACP ball round. It left a big bulge but no exit (above right).

Complete penetration of loaded magazine with the Czech 9x19mm steel-core ammunition (top left).

The entrance hole in the first magazine struck by the Czech steel-core 9x19mm round (top right).

Complete penetration of both loaded magazines with steel-core 9x19mm Czech ammunition (above left).

In the foreground is the exit hole made by Czech steel-core 9x19mm; the second hole is the entrance hole in the second magazine (above right).

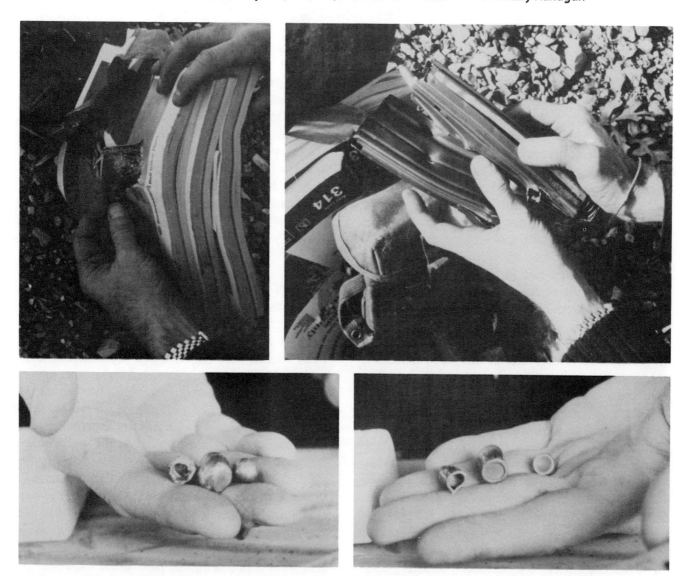

The Czech steel-core 9x19mm ammunition went through two fully loaded magazines, the pouch, and a 900-page telephone directory. The bullets were found in the dirt beyond the telephone books (top left).

A comparison of the dents in the second magazine: (top) Czech steel-core round completely penetrated both magazines; (center) 9mm Makarov left deep dent but no exit from first magazine; (right) .45 ACP ball penetrated first magazine but caught in second (top right).

A comparison of the recovered bullets after shooting into loaded magazines: (left) 9mm Czech steel-core; (center) .45 ACP ball; (right) 9mm Makarov (above left).

A comparison of the rear of test bullets after recovery from the magazines: (left) 9mm Makarov; (center) .45 ACP ball; (right) Czech steel-core 9x19mm (above right).

Additionally, the military cartridge, except in very rare cases involving intelligence operations, will have to comply with the requirements of the Hague Conference of 1907. Although not wishing to belabor the point, the main consequence of that compliance is to effectively ban expanding or exploding ammunition from the military handgun. Certain counterterrorist organizations may use expanding ammunition when they perform their antiterrorist roles; the theory is that terrorists fall outside the Conference conventions, for the most part. This does not mean that individuals will not use more effective expanding ammunition. (I remember an individual using 150-grain, soft-point ammunition in his .308 rifle in Vietnam.) It does mean that no government will issue such ammunition or sanction its use. Similarly, merely banning expanding ammunition does not mean that more effective ammunition cannot be developed or issued that avoids the restrictions while still giving the soldier a round more effective than conventional military-style handgun projectiles. The French THV rounds come immediately to mind, as do the German BAT rounds. What all of this means, ultimately, is that the common response in civilian circles of adopting an expanding round to be used against an unarmored or soft target cannot be slavishly followed in the military.

Additionally, the military procurement officer will view the handgun as an item of very low priority in comparison to artillery rounds, land mines, mortar rounds, and heavy and light machine guns, not to mention the service rifle. The procurement specialist will be concerned about the cost of manufacture, use of raw materials, and shipping expense and bulk. Hence, all things considered, the smaller and lighter, the happier the procurement specialist.

Similarly, if the ammunition can be used in another item in the military arsenal, so much the better. This fact explains why the 9x19mm caliber became so popular. The widespread development of the submachine gun and its appetite for ammunition meant that any military handgun was very likely to be of the same caliber. With the general disappearance of the submachine gun from military inventories, this may no longer

be so pressing. Because it was the case for so long and so much 9x19mm ammunition is available in the inventory, however, much of the development of modern handguns has taken place in envelopes for this cartridge (i.e., the Glock and SIG series of pistols). As Jeff Cooper put it so well, people shoot 9mm pistols for the pistol not the cartridge. Perhaps the next 50 years will see a trend away from cartridge adoption following the submachine gun cartridge. One can only hope so, because cartridges that fare poorly in submachine guns perform superbly in military handguns.

THE LANYARD

Another interesting issue when discussing military handguns is the use of the lanyard. The civilian gun carrier rarely uses a lanyard, and even among law enforcement departments, they are usually found in those that have maintained very old-fashioned uniforms where the lanyard is a traditional part of the attire (e.g., the Royal Canadian Mounted Police) or by the officers on horseback. Yet, while unusual in the nonmilitary context, a lanyard makes a lot of sense on the military handgun. The average infantryman will encounter much more rigorous activity than any law enforcement officer. He typically will be operating in a rural environment where a pistol could easily be lost or in the air. Even if none of this were a problem, just having the capability to fasten the weapon to the wrist while sleeping makes the lanyard a very practical item for the military person. Naturally, care must be taken to avoid the lanyard coming between the hammer and frame of the weapon and tying it up, thus rendering it impossible to fire (a problem known to have caused at least one officer's death during World War I, according to Captain Tracy). But even with this risk, the great benefits of the lanyard are enough to require it be present on the military handgun.

THE HOLSTER

Holsters are another factor that must be carefully considered when discussing the military handgun. Military holsters are different than the holsters used for law enforcement or civilian needs. First, we have the problem of procurement: no procurement specialist would tolerate the horsehide holster selected by the civilian. The military wants a holster than can be quickly manufactured out of low-cost material, will last for a long time in environmentally hostile situations, and will not soak up gas, etc. Yet, while wishing to protect the weapon (realizing as we do that the weapon will be in a very hostile environment), we must not fall into the European trap of making a holster into a pistol case. The weapon, if it is to be used, must be available rapidly. I do not think anyone can draw a pistol quickly from a German World War II-style Luger holster.

Another difficulty is in determining where to carry the military handgun. Obviously, a staff person can simply carry the weapon on the right hip easily enough. But we are concerned about the fighting man. With the typical load of spare magazine pouches found on belts around the middle, I think it is very unlikely that a right hip holster will work for our warrior. By the time you put four to six magazine pouches on a belt, then add a canteen or two, a few grenades, and your rucksack, all over your body armor, you are not going to be able to wear a hip holster. At least, I cannot do it. The only satisfactory alternative seems to be a shoulder rig, chest carry, or a breast pocket.

The shoulder rig is not very comfortable in hot weather and interferes with the rucksack, and for that reason I discount it. The chest rig works much better. Naturally, the weapon is exposed to mud when you are crawling through it, as you will be doing commonly, but a proper holster will keep much of it off the weapon. Another real drawback is that the weapon is out of use. I believe this risk is worth the convenience of this method of carry.

Last, simply sticking the weapon in the upper breast pocket is not bad. I used this method with a Smith & Wesson M60 .38 Special revolver with great effect in the 1968-77 period when carrying an M14 or M16A1 rifle. One advantage is that no one knows you are carrying the handgun, and that alone can come in handy. It also avoids the cost of a holster. The disadvantage is that it limits you to a very small, light weapon, while the chest carry allows you to carry a serious fighting handgun, such as the Glock 17, quite easily.

Other common ways to carry weapons in the civilian context are simply not very functional for the infantryman. Carrying your weapon on your ankle will quickly get it wet and muddy, as any infantryman knows. Your pants pockets are already bulging, and those magazine pouches containing very valuable rifle ammunition preclude the handgun's being carried on your belt line. Frankly, 90 rounds of rifle ammo are more important than the military handgun, so replacing a magazine pouch with the handgun does not make tactical sense.

WHAT TO LOOK FOR IN A MILITARY HANDGUN

There are a lot of military handguns around, and many fine weapons have been developed since the end of World War II. What are really the most—and least—desirable features for the military handgun? We can dispense quickly with the obvious requirements of reliability, controllability, and stopping power, and concentrate on the less obvious.

Reliability, Controllability, and Stopping Power

Any weapon to be useful must be reliable. Although reliability, at least among standard brands, is commonplace today, such was not always the case, especially in the early days of the semiautomatic pistol. For the weapons expert who has been shooting the combat handguns for 30 years or more (like me), controllability is a different issue than it is for a 19-year-old enlistee who comes to the military with little knowledge or exposure to weapons. Stopping power is likewise frequently different. In the civilian or police context, you are likely to encounter individuals who are using painkilling drugs or alcohol and who must be met with the minimal amount of deadly force to stop their attack. Along with your weapon, you must also equip yourself with the knowledge that severe criminal and civil consequences lie ahead if you err on the side of too much force. Yet these individuals, except in rare circumstances, are unlikely to be wearing armor or items that will stop handgun bullets, such as magazine pouches with loaded magazines, helmets, shovels, and all the other items of equipment common for infantry troops throughout the world. The military man or woman wants to *kill* his or her attacker; the civilian or police officer wants to *stop* the attacker. No one will second-guess the military man who fires (except in highly unusual situations), whereas the nonmilitary shooter can expect a very complete and thorough review of his actions. Therein lies one of the primary differences between the selection of a military and civilian weapons.

Finish

The weapon should be durable. With the advent of rust-resistant finishes, it makes absolutely no sense to accept something less. The weapon can be made of stainless steel or have a rust-resistant finish such as Teflon applied. The military man has enough on his mind, and any time saved by not having to be careful about finishes is well rewarded. It would also be nice if the weapon had a dark finish to avoid reflection problems in the field.

Night Sights

Night sights made of radioactive elements are not very useful in the civilian context because they are rarely bright enough for the dim light of civilian shooting. In those situations you must make certain of your target before firing and ensure that you are exercising proper deadly force. For military situations, such restrictions do not apply, and hence the night sights are much more useful. Also, the military shooter is likely to use the handgun even more frequently than the civilian in total darkness, thus adding another element to its usefulness.

Trigger Guard

Since the soldier may be outdoors in extremely cold weather, an oversized trigger guard is helpful. It allows the insertion of the finger into the trigger guard even while wearing gloves or mittens. Probably, the soldier will not be sitting in a warm patrol car as the police officer typically does.

Safeties

It should also be easy to put the weapon into a condition to fire. Thus, safeties should be few in number and easy to disengage. Keep in mind also that the typical military shooter will not be nearly as well trained as the civilian, hence the "KISS" principle is critical. However, the anti-snatch concerns of the civilian shooter are notably absent with military handguns.

Magazine Capacity

All other things being equal, a high magazine capacity is useful. You can avoid carrying spare magazines, leaving you extra room on your belt, and still have additional rounds to deal with multiple foes or the event that your first rounds encounter objects that impede their penetration of the body. However, the weapon should have a small grip, especially since the weapon is likely to be used by females or small-statured people.

Shape and Size

The weapon should also be free of sharp edges that cut the hand or cause the body to be injured. The lighter the weapon, the better, for our trooper is already burdened with enough equipment. The smaller the weapon, the better, simply because it gives the soldier more places to keep the weapon when in the field.

Maintenance

From the procurement standpoint, the parts should be easily replaced without tools and naturally should be interchangeable. The individual soldier may not care about this feature, but those charged with adopting a general military weapon for units certainly are legitimately concerned about such issues.

Accuracy

The weapon needs to be accurate enough for the needs of the soldier, but it does not have to be a target weapon. It should be accurate enough to allow a well-trained shooter to keep all his rounds on a man target at 25 yards at the least. The caliber should be sufficiently powerful to disable an individual if struck in a critical zone, but the problems about ammunition selection and needs of penetration must be recalled. Certain calibers, however, clearly fail this test and can be discarded at once.

The Uniqueness of a Military Weapon

Lastly, keep in mind the unique needs and uses of the military handgun. Do not fall into thinking that civilian or law enforcement experiences and needs can be used to determine the proper military handgun. They cannot, any more than military considerations can be used in the civilian or law enforcement arena. Each field has its own dynamics, and it is critical that these differences be recognized so that a proper handgun can be selected.

The Good, the Bad, and the Surprising Military Handguns

The ordnance officer is concerned about picking a weapon that can be produced quickly with low demands on skilled time and valuable resources. He demands that the parts can be interchanged at the forward edge of the battle area and that the weapon will last long enough to meet the expected need, which actually may be very few actual rounds fired through it.

The conditions found by the infantryman at the front are always much worse than those found by a police officer or typical civilian. Thus, the exposure to the elements will be greater for the military weapon. On the other hand, the military weapon will be in the hands of soldiers who are generally being supervised by others, and they are only actually using their weapons for a very brief time in their careers. The civilian or police officer, however, will be carrying the weapon for long periods under a wide variety of circumstances and not be supervised.

Mistakes in a military situation, while always tragic, are not nearly as involved as in a nonmilitary situation. In a military environment, there are so many ways to get hurt or killed that an accidental discharge from a handgun is almost meaningless. One's whole world is dangerous, whereas in a nonmilitary situation, such accidents are of much greater importance.

For all of these reasons, and more, which are apparent to anyone who has given the subject much thought and who has served as a military officer, law enforcement officer, and civilian gun carrier (as have I), the selection of the proper weapon for a military unit may well be different from that selected for other purposes. When you first approach this subject, you may not realize this; upon reflection, it becomes obvious.

The line of thinking does not go just to the selection of the weapon, but also to the caliber. What may make good sense in a law enforcement or civilian situation may not be a good selection for the military handgun. Obviously the different concerns mentioned earlier, which may lead us to a different conclusion when selecting the military handgun, may likewise cause us to select different calibers, even if we have the luxury of selecting a caliber from scratch, which is not typically the case.

Last, when selecting a military handgun, we must ask, "Why are we really selecting a handgun?" In a real military unit, you fight with artillery pieces, machine guns, grenades, mines, rifles, and air strikes, and very little is actually done with handguns. This does not mean that they are unimportant; it merely means that it is rare to actually use the handgun in combat. Of course, when you use it, you really need it. But we must decide if we are selecting a weapon to be carried by the individual military man on the front. That was the approach taken by the U. S. Army during World War I when the goal was to equip every U.S. soldier with a handgun.

As a former company-grade infantry soldier, I think this is an excellent idea. I know I would much rather have a 5-shot .38 alloy-frame revolver and a Boy Scout knife than the commonly seen big fighting knife or bayonet. I could use the folding knife for all my cutting needs, kill better with my revolver, and use my entrenching tool instead of a knife for any attack purposes, and still dig my holes to hide in. Are we selecting a weapon for the pilot, staff officer, rear-echelon trooper, or criminal investigator? Those people do not carry a rifle, so if they need to fight they will use their handgun as a primary fighting tool. Also, such people are, frankly speaking, more valuable to the army than the front-line solder and, hence, a greater expense in protecting them can be accepted—both in acquisition costs and training.

Faced with these dissimilar needs, we can either select different weapons or attempt to somehow select some weapon that will be acceptable to both groups of basic users, realizing at the same time that we are destined to fail in some measure at either extreme.

After all this, you're probably wondering what combat handguns are at the top and bottom of my list. After all, I have tested most of the main military-style weapons in use for the last 125 years. With the understanding that my heart is with the infantry soldier who will be carrying his or her handgun in addition to the shoulder weapon, I can recommend the following handguns as the best possible weapons.

The author's M65 357 Magnum custom revolver with the barrel and butt both chopped, shown here with THV .357 Magnum ammunition.

THE BEST COMBAT HANDGUNS

1. *Smith & Wesson M65 .357 Magnum,* with the barrel trimmed to 2 inches and the butt trimmed, loaded with Hague Conference-approved THV ammunition. This is my personal favorite, although it is only available as a custom creation. This is the weapon I selected after all my testing and evaluation as my military weapon for the future.

2. *M940 Smith & Wesson.* I think this comes very close to the M65 when loaded with THV ammunition.

3. *Glock 17.* It is a tough, lightweight, high-capacity weapon whose cartridge has good penetration. Although I have some real doubts about the use of the Glock as a law-enforcement weapon (because of problems associated with weapon

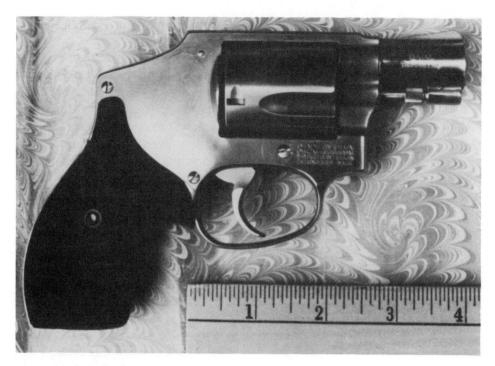

Smith & Wesson M940 9x19mm.

snatch and holding suspects at gunpoint), it is the best choice for a military handgun.

4. *SIG P225 in 9mm or the SIG P220 in 45 ACP.* Both are lightweight, extremely reliable, accurate, and easy to use. As with the Glock, they may not be suitable for law enforcement tasks because of the lack of a manual safety, which is often helpful when dealing with weapon snatches and the problem with holding suspects with a cocked pistol.

5. *Smith & Wesson .38 Special 2-inch stainless-steel Chief Special or the M642.* The Smith & Wesson .38 Special was much sought after during the Vietnam War, and the M642, its successor today, is even lighter to carry. Either makes an excellent companion when thrust in the top pocket of your uniform jacket.

THE WORST COMBAT HANDGUNS

1. *Dreyse .32.* This must clearly be considered the worst pistol tested.
2. *"Brixia."* This Italian handgun was a close runner-up.
3. The *Montenegrin* revolver would be next in

line for this dubious honor.
4. *Japanese Type 26*
5. *Italian 10.35mm.*

Although any of these weapons would be better than the best knife in the world, they are all terrible examples of the gunmaker's and designer's art, especially when we consider what was currently available from other sources.

THE MOST SURPRISING WEAPONS TESTED

In this category are those weapons with which I was unfamiliar or those I would have thought only average or worse when, in fact, they turned out to be superb fighting handguns, not good enough to make the final top five but really great and within the top five in their own times.

1. *French M1873 11mm revolver.* Although not as powerful as the Colt 45, the M1873 is so far superior as a fighting handgun that there is simply no comparison. It was designed by people who knew what it took to make a serious combat revolver.
2. *Mauser M96.* When viewed on the formal range, it seems clunky and hard to handle, but on the cinema range, you suddenly realize it is a real fighting pistol. The men who used this pistol had a very serious fighting handgun. It is fast to use, accurate, and powerful. The M96 is a good gun that only reveals its qualities in the hands of those who are sophisticated enough to use it as it should be used: as a fighting tool, not a remote-control paper punch.
3. *Roth-Steyr M1907.* This is perhaps the real jewel of the entire test series. This is a weapon that few get to shoot, and you often read many disparaging comments about it in

the gun press. You add a detachable magazine and chamber it for 9mm, and you have a 1907-made Glock pistol. This pistol is so far superior to any other fighting pistol of the time as to be beyond comparison. A truly wonderful pistol!

4. *Webley Mark VI.* The Mark VI went into production in 1915 and was considered by many the ultimate Webley. Although not as well finished as the semicustom Webley-Wilkinson or Webley-Greene, this Webley—like the others—is a real combat revolver. It was not designed to shoot targets or game but rather to rapidly stop aggressive foes at short range. For this purpose, it is far superior to the Colt or Smith & Wesson of the period. Many who remember the $10 Webley pistols for sale in the 1960s tend to view them as junky top-breaks; instead, they are perhaps the finest combat revolver ever.

5. *Soviet PSM in 5.45mm.* When you first hear this handgun described, you can only wonder why it was designed and produced. After testing it, you realize that the Russians who designed it are the same people who routinely produce chess wizards. Its small size, lightweight ammo, easy carry, and ability to penetrate the armor now commonly worn on the battlefield are all clues to its mysterious origins, as well as traits that make it one of the best combat pistols of the late twentieth century.

• • •

It was great fun researching all the handguns for this book, getting my hands on them, and test-firing them for this book. As I stated in the Acknowledgements, the gathering together of more than 100 of the world's best and most exotic weapons for such a task could only have taken place in the United States. I trust this book will be a fitting memorial to those honorable men who created the Bill of Rights that separates the United States from so many other countries. Fifty years from now, it may not be possible to assemble and test such weapons because of their rarity or the prohibition against having such weapons. Even now, getting some of the weapons was a formidable challenge, as was tracking down the ammunition. I am certain that it will be more difficult in the future.

Even though a lot of work, time, and expense have gone into the production of this book, I am glad I did it. It allowed me to do something that no one else has been able to do (to my knowledge) and something that is unlikely to be repeated in the future. For me, it was an exciting, educational experience. For readers, my hope is that you have learned something that may someday save your life in some faraway outpost, educated yourself as to the possibilities of that "old pistol" in your collection, or, if nothing else, acquired greater respect for our predecessors who waged war on each other in generations past with these war weapons.

Preparing Your Handgun for Combat Duty

One of the key objectives of this book is to prepare you in the event you wind in some distant backwater with only an odd or unusual weapon available to you. You need to know what to look for in a combat weapon, as well as how to handle a weapon that you may never have encountered before and what to expect of it. Bearing in mind that when you are in such wastelands (be it Trenton, New Jersey, or Asia Minor), any pistol is better than no pistol.

Let us assume that you have now found the handgun, whether from the local bazaar, black market, or a captured weapons cache. What should you do next? How should you prepare?

CLEANING AND INSPECTING

First, make certain the weapon is not loaded. You do not want to get shot in the eyeball because of your ignorance! Assuming the weapon is not loaded, you should clean the weapon properly. Weapons bought or found under questionable circumstances are frequently dirty and caked with grease. Assuming that typical gun-cleaning equipment is not available, gasoline, diesel fuel, or even soap and hot water will suffice. Use a toothbrush to clean the handgun carefully, removing old oil and grease that

may tie it up. A handkerchief can be used as a bore brush if you do not have a cleaning rod, or, if you cannot get a brush, a little window screen material wrapped about a rod (an ink pen filler, for example) will do a fair job.

As you clean the weapon, examine it carefully for cracks in the metal. Look at the firing pin and see whether it is broken. Test the action—does it fall sharply and hit with a snap, or is it sluggish? If the springs are tired, you can stretch them or put a shim in to help out. Then reassemble the weapon. Keep in mind that you rarely need to use force. Do not hammer on the parts to take it apart or reassemble. Look at the weapon and think about it. All weapons are designed by engineers; some for better, some for worse, but none of them are designed to be broken apart with a hammer.

REASSEMBLING AND LUBRICATING

After you have carefully cleaned the weapon and examined all the parts for cracks and weak springs, it is time to reassemble and lubricate. Absent gun oil, you can use sewing machine oil or even oil from your car. I had a Colt M1911 freeze up once from being totally dry, and a simple drop from my dipstick got me back in business. A friend in SAS reports that cleaning the weapon in gin works in the desert because the alcohol cleans the weapon yet the berries from which the gin was made contain enough oil to provide adequate lubrication without the risk of sand clogging the weapon. Keep in mind when you lubricate that you are not storing the weapon. Light lubrication only!

TESTING OR IMPROVISING AMMUNITION

Next look at your ammunition carefully. It is unlikely that you will get factory-fresh ammunition out of the box, but we can all hope. Assuming you have only a limited number of rounds that aren't straight from the factory, how should you approach the matter? First, check the ammunition for corrosion. Wipe the cases carefully and examine for cracked cases and high or loose primers. Then run the ammunition through the weapon to make certain the weapon will function with it. Check all rounds in the cylinder of a revolver and function the weapon manually through the magazine if it's a self-loader.

Assuming all is in order, you then have to test the weapon. If you have plenty of ammunition and can test the weapon openly, everything is easy. A more typical scenario is that you will only have a cylinder or magazine full (if that) of ammunition. Still, you must know whether it will work, so even if you have only five rounds, you've got to sacrifice one. Lacking better facilities, fire the weapon into telephone directories. Turn up the volume on the radio or TV, run a car, or do something to mask the sound.

Even in a big city, you can usually get away with one round being fired. People might hear it but cannot be certain about what they heard or where it came from. If that is too risky, pull a bullet and pop the primer off. Put the weapon against a book or telephone directory, put a pillow or clothing over the weapon, and even the noise of the primer should be masked sufficiently to avoid even the sharpest of ears. Then check the primer to see if it is hit, centered, and properly dented. If you pulled the bullet, look at the powder. Does it look like it is decomposing? Put it in an ashtray and light it. Does it burn with an even flame?

Naturally, testing your weapon this way will not tell you everything, but at least you will know that the weapon and at least one round of ammunition worked all right. Statistically, you are in good shape now. Do not forget to clean your weapon again since you must assume that the primer is corrosive—unless you know that is not the case.

Last, you may be in a position where you can get a weapon but not the proper ammunition. Take the weapon and hope that you can find the ammunition later. Even if you cannot get the proper ammunition, you may be able to modify what you have to fit. For instance, you can expand the base of a .32 ACP to make a rim, and it will work in an 8mm M1892 revolver. The .380 ACP pistols can fire .32 ACP and .38 Super. You can shoot .380 ACP ammo in a 9x19mm pistol. You can fire .38/.40 in a .44/.40 pistol by wrapping paper around the cartridge. You can single-load .30 Mauser into a 9x19mm pistol and fire it, and .30 Luger will work fine, also. Be creative. Your improvised round may not be ideal, but it will work until something better comes along.

• • •

The first rule of gunfighting is to have a gun. But you frequently may have to do with something less than your ideal. Although I would prefer to have my Glock 17 9mm loaded with THV ammunition in a crisis, it is not likely to be stocked at the corner store in northern Asia. But even the worst possible handgun—that is, a .32 Dreyse with four questionable rounds (but tested as noted)—is better than a stick.

About the Author

Author Timothy John Mullin graduated magna cum laude from St. Louis University and was later awarded his juris doctor degree from the University of Chicago Law School. Mr. Mullin served seven years with the U.S. Army, first as an infantry officer and later as an officer with the Judge Advocate General's Corp. Upon leaving the army, he was awarded the Meritorious Service Medal in recognition of his services.

During the years 1976 and 1977, Mr. Mullin served as Chief of the Department of Defense Police, St. Louis area, and as a Deputy U.S. Marshall. Since that time, he has engaged in private law practice. For the past 17 years, the author has also served as training officer for a local Missouri police agency, establishing a modern firearms training program that emphasizes legal as well as tactical aspects.

Mr. Mullin is the author of the well-respected book *Training the Gunfighter*, which deals with firearms training for police agencies, and has written numerous articles for various firearms and police-oriented magazines in the United States, the United Kingdom, and Belgium.